HAPPY TO FLY

ALSO BY ANN WELCH

Accidents Happen
Pilots' Weather
The Story of Gliding
Hang Glider Pilot (*with Gerry Breen*)
Soaring Hang Gliders (*with Roy Hill*)
New Soaring Pilot (*with Lorne Welch and Frank Irving*)

HAPPY TO FLY

An Autobiography

Ann Welch

JOHN MURRAY

To all my Family

© Ann Welch 1983

First published 1983
by John Murray (Publishers) Ltd
50 Albemarle Street, London W1X 4BD

All rights reserved
Unauthorised duplication
contravenes applicable laws

Typeset by Inforum Ltd, Portsmouth
Printed and bound in Great Britain
by the Pitman Press, Bath

British Library Cataloguing in Publication Data
Welch, Ann
  Happy to fly
  1. Welch, Ann   2. Air pilots—Biography
  I. Title
  629.13'092'4   TL540.W/
  ISBN 0–7195–4033–X

Contents

	Introduction	1
1	Growing up	6
2	The good west wind	18
3	Downhill, uphill	33
4	Early days in ATA	40
5	Spitfire	56
6	Peace freedoms – and problems	69
7	High hopes in the Alps	83
8	Redhill – into high gear	91
9	We win at last	96
10	No dull moments	103
11	Trailer days – and nights	121
12	A little flying on my own	135
13	No pesetas in the pampas	152
14	A Worlds of our own	164
15	A different sort of competition	172
16	An empty corner of Texas	180
17	Change is nothing new	187
18	Honorary temporary Australian	199
19	Kössen and Canada	210
20	Loss, but many compensations	221
21	Messing about with microlights	229
22	The dream remembered	236
	Index	243

Illustrations

BETWEEN PAGES 88 AND 89

1. AW with Charlie Allen and his Puss Moth, 1932
2. AW's Royal Enfield motorbike, 1933
3. Box Brownie Bulldogs at Biggin Hill, 1933
4. G-AAIM, AW and Instructor Nash, 1934
5. Brooklands between the wars
6. Anglo-German camp at Dunstable, 1937
7. Dunstable, 1937: AW splicing retrieve winch rope
8. Skiing above Berchtesgaden, 1938: falling in the snow
9. Rossfeld ski hut, 1938
10. AW and Grunau at Dunstable with dog, Bogie, 1938
11. Surrey Gliding Club, 1938: AW repairing Grunau
12. John Sproule instructing Arthur Saville on the Dagling
13. ATA: AW and Spitfire at Hamble, 1942
14. Torcross, summer 1947: AW with Vivien and Liz
15. Supper at the Mynd, 1946
16. VJ on the hilltop at Salzgitter, Germany, 1946
17. Minimoa after Silver C distance to Lüthorst
18. Surrey Club Weihe landing at Redhill, 1947
19. At Redhill with Lorne and prototype Olympia, 1947
20. Surrey Club skiers at Zermatt, 1948
21. Swiss Moswey above mountains near Samedan, 1948
22. Jan, 22 years before qualifying as a doctor

BETWEEN PAGES 184 AND 185

23. British Team for World Championships, Spain, 1952
24. Glider retrieving in small French town, St Yan, 1956
25. In Horse Guards Parade, after tea with Prince Philip
26. Nick Goodhart landing at Leszno in the Skylark 3

Illustrations

27 Nationals at Lasham; Prince Philip and Philip Wills
28 AW about to fly Philip Wills's Skylark 4 at Lasham
29 Retrieve winch flat car at Zar in the Tatra
30 Irene Zabiello and Adam Dziurzynski, Zar, 1962
31 British Team at Junin, Argentina, 1963
32 First take-off at Junin; start of the dust fog
33 Hours of waving from the Citroën 2-cv in Junin
34 Our World Championships at South Cerney, 1965
35 South Cerney: preparing for take-off
36 Our 'village square' at South Cerney
37 The T49 Capstan two-seater during trials at Lasham
38 Testing the Skylark 2 at Lasham
39 Wally the Met, Ingo Renner and C.J. Ridley, Vrsac
40 George Lee, Great Britain
41 Weighing a competitor before take-off
42 France, 1979: World Hang Gliding Championships
43 Japan, 1981: AW as Chairman of the Jury at Beppu

SOURCES OF ILLUSTRATIONS

1, 2, 3, 6–10, 12, 14–17, 20, 22, 24, 25, 27, 29, 30, 32, 33, 35, 36, 39–42: the author; 4: R.L. Knight, Barnstaple; 5, 21, 37: *Flight International*, London; 11, 23: Fox Photos Ltd, London; 13: *Vogue* © The Condé Nast Publications Ltd; 18: Graphic Photo Union, London; 19: Charles E. Brown; 26: Bernard J. Koszewski, Warsaw; 28: PA-Reuter Photos Ltd, London; 38: Lawrence Hanley, 43: Noel Whittall.

The drawings and engravings are by the author; those on pages 15 and 21 are reproduced from *Aeroplane*; that on page 17 from *Pilots Only* (1936); and on page 76 from *Sailplane and Gliding*. Other illustrations are by R. Roux (page 121) and Peter Fuller for *Sailplane and Gliding* (page 152).

Introduction

Suddenly I was sixty. Then almost at once, it seemed, I was sixty-five. What was happening to those years which had once seemed to stretch far into the future, filled with so many marvellous things to do and still to be done? Now, here I was, helping to teach people little older than my granddaughter to become hang gliding and microlight instructors; though after only a few not very successful flights I had realised that, for me, the delights of hang gliding had sadly come twenty years too late. I must be mad; I should be pottering in the garden or going to boring coffee mornings, not running round without a moment to spare.

Once, a long time ago, I knew that all I ever wanted was to fly, to just float around in the cool sky, playing with the clouds, looking at the land and sea below, toylike and beautiful. Or, when I had to be on the ground, to create things: to paint pictures, to write about flying, or make things in wood which smelt so good. But here I was, involved in paperwork and those faceless international organisations which years back I had thought so unimportant – I had all the ideas I needed of my own. What was it that had determined the way I would, or should, go? What decisions had I unwittingly taken that caused the linking of events which had inexorably led from those lovely dreams of flying to the administrative pressures of today – but also to the fascinations of an aviation far beyond my early imagination. My dreams were just of being in the air like a bird, and did not include travelling comfortably to the Pacific in a day, or 'flying' the 747 simulator in Seattle; this felt so real that I believed landing from a cockpit 30 ft up would be difficult, but it was not. Certainly I never had any intention to become what I am or to do many of the things that I have done; the possibilities were not even known to me. So how did it happen?

In 1903 an organisation had been born called the Fédération Aéronautique Internationale (FAI) for 'the encouragement and control of all sporting flying', but I did not really know anything about it until I had been flying almost thirty years. It was in 1965 that I went to one of its meetings in Paris to discuss rules for the World Gliding Championships in England, of which I was to be the Director. Afterwards I was invited to Paris again to consider future rules and found myself becoming increasingly involved with the FAI Gliding Committee, CVSM (now CIVV – Commission Internationale de Vol à Voile), and after that as British FAI delegate, a Vice-President, and Editor of the annual *FAI Bulletin*. I liked the FAI because it was a great and civilised gathering of enthusiasts, people from all over the world working together for the sort of flying we loved; and because it also gave me those undreamt of opportunities to see so many countries – Chile, Iran, New Zealand, and Japan – and to be with my flying friends.

My involvement with the FAI came because of organising a World Championships, but being put in charge of that had resulted from my having been manager of the British gliding team over the previous fifteen years; so I was supposed to know something about the problems. Those years had taken me on remarkable expeditions, some hilarious, a few solemn, but all different; and had brought the realisation that the fun of flying was not only being in the air, but being together with people whose sense of humour was the same as mine – well, often it was.

But becoming team manager in the first place, in 1948,

was due to the happy chance of being in the right place at the right time; simply because I was on the British Gliding Association Council representing the Surrey Gliding Club. And I had come on to that, not because I liked administration but because I had found that I really enjoyed teaching flying. It gave my creative energies plenty of scope in trying to improve techniques and methods; which were much needed if my own experience as an instructor at that time was anything to go by! And I had become an instructor for no better reason than spending every spare moment out on Dunstable's windy and beautiful hilltop in 1937, and was one day seen with my hands in my pockets – actually feeling if there was enough money there for another flight. With no other instructor present someone told me to help the newcomers – those unfortunates who had started about two months after I had – to fly from the top to the bottom of the hill without reducing the primary glider to matchwood. It was a challenge. I liked trying to find the right words to say so that the beginner understood what he was supposed to do, and it was satisfying if each time he did it better. To do better myself I listened more carefully to the chief instructor, Tim Hervey, to learn how he did it.

Before that I had wanted only to fly, like some disembodied spirit far from reality, though the dream to me was very real. And I had been able to fly only because so many people had helped me change my dreams into hours in the air. All sorts of people, some unknowingly, had given me something, from words of encouragement to flights in the back seats of their aeroplanes.

So this book is not only about the pleasures of flying, but the places to which it has taken me and the often ludicrous situations in which I have found myself, and the friends without which life would be no fun at all. It is, I suppose, a thoroughly one-sided history of gliding championships and what I believe were perhaps the best years ever for what is now euphemistically called recreational flying; carefree years full of new and exciting adventures – not all of which worked out – and of being a part of those years. My flying has never been dramatic or ambitious; no fighter pilot stuff or long pioneering flights to Australia. It has just been fun, sometimes a little frightening, but I would not have missed any of it. But I am still not sure what it was, back in the beginning, that started me off on such a single-minded journey; or what it is, when there is great freedom of choice, that produces an overwhelming desire to do only one thing. I think it is only rarely that an isolated happening determines a whole future, it is probably more of a trigger mechanism. For me it was an unexpected five-minute flight that somehow collected together all the loose ends of interests and ideas which had until that moment no particular direction.

It was during the last summer holiday, of many that my family spent on the wild and empty north coast of the Cornwall of my forebears, when Alan Cobham's Circus came to a small grass field near Wadebridge. I was taken to see it – I am not sure how we went, as few people had cars – and was immediately captivated by the row of brightly coloured aeroplanes with their unidentifiable smell; and by touching one for the first time. All around there were banners draped on the hedges and more people than I had ever seen together in Cornwall. One after the other aeroplanes rumbled into the air, flew round, and bounced back on to the rough ground; and when without warning I was given the chance of a flight I became totally unaware of anything else around me. It was in a rather ugly, squat biplane with three engines, called an Airspeed Ferry; but I did not mind what it looked like. Jumping in as soon as I could I found a seat with an unobstructed view – not easy with all those struts and engines – and waited in breathless anticipation. It was not a long wait; Cobham's pilots knew how to keep moving. We bumped and rattled over the grass. Then it was all smooth as the small fields and stony Cornish hedges fell away at an angle changing, as we turned, into endless miles of glistening sunlit sand with a rugged headland meeting a deep blue sea. It was a new world in which I did not even recognise the same Camel River Estuary and Pentire head that I knew and loved. Then farmland was underneath and the engine noises died away. I knew we had landed only because with my face pressed against the glass my chin bumped on the window frame. The door was opened and we clambered out, me last, slowly, reluctant to leave. I was a small, skinny and tousle-headed thirteen-year-old, the year was 1930, and I wanted to fly.

AIRLINER 1928

1 · Growing up

Born in London among the infrequent Zeppelin raids of the First World War, my first few years taught me little of use to flying, though I suppose my first lesson in stability was when, aged four, I joined my small brother, John, in the open bottom drawer of a huge wardrobe which promptly fell on its face – fortunately the drawer jammed open; and I may have learnt something about structures when the front door slammed in the wind locking out everyone except me. Through the letter box I was anxiously instructed to get the key from the high shelf out of our reach; I am not sure how structurally sound was my pyramid of chairs and wastepaper baskets, but I made it to the key.

Shortly after that we moved to the country – it was country then – staying in a hotel while our house at Bickley in Kent was built, and for the first time I could see that the sky was big, and full of beautiful moving clouds which I could watch all day; dark rain would pass leaving a tracery of white feathers against a delicate blue, and in the evening thousands of little cloudlets grew golden then fiery red. Grownups said it was a mackerel sky, but I did not know what a mackerel was; it was the light and the air which I loved, and the wind on my face. Even more fascinating were the aeroplanes that droned across this sky, sometimes very low over my head. I did not know that they were the first airliners, pioneering their way from Brussels, Amsterdam and Paris to Croydon. Looking at a map many years later I found that the Paris aeroplanes should have been much further south, but in rain and storm must have been following the low ground, creeping along the railway line to arrive at Croydon from the north-east. Certainly in such weather they flew low, sometimes turning around trees before disappearing into the wispy murk. They did not come at night, though occasionally cabin lights moved slowly through the dusk. In summer they flew high and serene, clear against the sky. I was fascinated by these aeroplanes, but they were remote from me and I never thought of flying them.

In 1924 we moved into our new house, I going to school all week and my parents happily gardening all weekend. I was interested in neither, longing only for our summer holidays in Cornwall; for those clean washed days when grey curtains of rain gave way to a sky of brilliant blue, over an even deeper blue sea with the whitest of roaring surf. On such days I would lie alone in the sand dunes, warm in the sun, and just look at the sky and the sea through the waving grass. Sometimes I would try to paint it in watercolours but could never achieve the brilliance of reality. Of course I also swam in the waves, explored rock pools, and ate strawberry jam sandwiches gritty with sand; and one day, aged five, I was given the helm of the Padstow to Rock ferry when it was an engineless sailing boat, and felt the wind through the tiller in my hand.

And so the years of my childhood went by. Except in school, time never dragged. I painted pictures, embroidered table mats, made a duplicating machine which would produce about twenty smudgy purple copies out of a foul smelling jelly; and a projector from a biscuit tin and a magnifying glass which showed postcards – you had to put the pictures in upside down – and jigsaw puzzles with a fretsaw. I did not see a lot of my brother; in Cornwall he was happy all day at Trenain farm where we stayed and at home he disappeared to boarding school. My own school work was unimpressive as I spent my time looking out of the window. Then I had that flight in Cornwall, and soon after I saw a scale model aircraft kit in a shop window. It was about the first – rough chunks of wood in a box and called a Hawker Fury; which I knew about as they sometimes flew over our house from Biggin Hill. It cost one shilling. I saved and bought it, opening it in the seclusion of my bedroom with a strange excitement I could not understand; and made it immediately.

It was when I was about fifteen that I met Henry Williamson, the author, in south Devon, at Torcross where my aunt lived; and I had never met anyone like him before. His mind was full of ideas new to me and he could invent stories so that time passed unnoticed. He visited us at home often but his unpredictability – disappearing in the middle of a meal or arriving through a window – caused consternation to my father, who was never sure whether to appreciate a genius or complain about bad manners. But Henry showed me peregrine falcons and gannets superbly flying from high north Devon cliffs, and did not laugh at me when I talked about aeroplanes or showed him my growing collection of models. I had soon run out of the few kits available and was building my aeroplanes from cigar boxes, cutting out the wings with my fretsaw, using florist's wire for the struts, and casting lead wheels in a tiny sand box. I drew the plans in a school arithmetic book to a scale of 6 ft to 1 in (1/72 as I discovered later when my sums had a little more practice), and I became quite good at drawing them just from photographs in magazines. In this way I built a model of every British light aeroplane made at the time – and there were a lot.

In 1932 the first issue of *Popular Flying* appeared on the railway station bookstall which I passed each day walking to school. I bought it and read everything in it, and realised that it was not just aeroplanes I liked. I desperately wanted to fly them. But I could not, I was still only fifteen, so I followed book instructions, practising for hours in my room with a poker in my right hand, a pencil for a throttle in my left and my feet on two boxes, doing take-offs, turns and landings; but I avoided stalling, which the book made to sound alarming and was not in the least clear what one should do about it.

miles was a long way to go. On 28 March G-AACI, CITY OF LIVERPOOL, was not so lucky when it crashed in Belgium killing the pilot L. Leleu. But on fine days there was no excuse except for the fun of flying for the majestic Heracles G-AAXD to float past, unhurried, just over our roof. Sometimes KLM flew low over the Kentish fields; one, PH-AIE, appeared in my diary no less than twenty-five times that year – it became quite a friend. Then the corrugated Junkers G-31 of the Lufthansa started coming, always on time, and later the incredible G-38 with windows in the wings. It flew very slowly, looking absolutely enormous and somehow sinister.

Most of the RAF aeroplanes I saw that year were the Bristol Bulldogs and Hawker Demons of 23 and 32 Squadrons from Biggin Hill, and when I discovered that the aerodrome was within cycling distance I wrote to the commanding officer to ask if I could come and look at them. I not only received back a kind invitation but a Flight Lieutenant Monty Whittle had volunteered – or been detailed – to escort me around and answer my questions. He did this with great patience, let me sit in the cockpits of these fabulous aeroplanes, took me to the mess for lunch, and then saw me back to my corporal-guarded bicycle. It was a visit which turned out to be the first of many as Monty asked me to dances at Biggin and often came to our house.

Now I found that I could also get to Croydon on my bicycle, which I had fitted with a larger back wheel, price 2s 6d, so as to be able to go faster without my feet having to rotate so rapidly. My visits took place at any time that I had 5s, the price of a flight; usually in a Puss Moth. Charlie Allen, the Air Taxis Pilot, was always kind enough not to hurry back on to the ground, and to let me lean over his shoulder and hold the stick. Then, one day, when no one was looking I went up into the control room in the tower. As I stood hesitantly in the doorway an important looking man – it must have been Jimmy Jeffs – asked me in and told me all about the aeroplanes he was expecting, and with him I watched Hannibal float by almost alongside the window.

Somewhat concerned about their daughter's strange and seemingly hazardous ideas my parents went in for diversionary tactics. One of these was to dispatch me to France for the summer holidays, to stay with the Graillot family in their ancient house surrounded by its own vineyards and apricot fields some 50 miles south of Chalons-sur-Saone. I had never been abroad before but soon came to love the hot, still days so different from the breezy English summer. Often we would walk through dry and scented fields noisy with grasshoppers to a quiet part of the river near Tournus to swim, and afterwards sit on its sandy shore eating hard black chocolate with crusty bread while the sun dried us. That was in the day; at night I dreamt of flying.

At the end of that year I left school, aged sixteen. The headmistress reckoned I would not get School Certificate unless I stopped gazing out of the window, and my father was tight on money keeping my brother at public school. But I think the final straw was when, without telling anyone, I bought a motorcycle for £5 so as to get more easily to Croydon and Biggin Hill. I rode it home uninsured and unlicensed — no one had told me about these things — and proceeded to go to school on it. Out of concern, because he was a very law-abiding man, my father dealt with the legal side of my new venture but was defeated by the displeasure of the headmistress. So I departed to the sound of her words at the end-of-term ceremony: 'Ann has been with us ten years.'

Still not reconciled to my ideas of flying my parents discussed the problem with Henry Williamson who suggested that I stay with Charles Tunnicliffe and his wife to learn about how to paint — more suitable for a girl. I had always liked painting and Charles Tunnicliffe's love for birds was after all a love of flying, so I did not feel hard done by, and they looked after me well. But it did not take long to discover that the Lancashire Aero Club at Woodford was only a bus ride away. The following Saturday I set off, arrived at the club gateway near a neat row of lovely aeroplanes, and wondered what to do next. I heard a voice asking 'Are you looking for someone?' and realised he was talking to me. I must have mumbled something about aeroplanes and he must have been kind because he showed me around the row of Avro Cadets and took me into the clubhouse, introducing me to members. One said he was just going up for a flight and would I like to come. My expression must have assured him for he soon found a helmet and goggles and a leather coat for me and installed me in the front cockpit. Incredibly, unbelievably, I was in the air again. Every weekend until I returned home I went to Woodford and each time I was taken flying by Colin Wilson, Peter Eckersley, George Youell, or Mollie Barnard. I have never forgotten them.

As I neared seventeen and the legal age at which I could fly solo I became increasingly worried by the clear and simple fact that I did not have any money to do so; and it was not at all easy then for a girl to get a job. So I continued to study the enticing flying club advertisements offering a complete A licence for £35 and equating it with others offering dual for £2 an hour and solo at £1 10s. There was also the problem of getting to and from whatever flying club it would be. To my great relief the transport problem unexpectedly became, I hoped, soluble. My parents, like most others at the time, did not run a car; but for some reason they suddenly decided to get one, ordering a new BSA with automatic transmission as my mother had never driven and my father not since the First World War. When the car arrived at the Bromley showrooms I, close on the heels of my seventeenth birthday, was sent off alone by bus to fetch it. Apart from my little motorcycling experience I had driven a car only three times – twice illegally in Henry Williamson's Silver Eagle Alvis with its crash gearbox along empty Devon lanes, and once in an MG Magnette for about half a mile down a private drive. I had never driven in a town. Luckily no drama occurred and I got the BSA home safely, on the strength of which I was told that I could now teach my mother! Between us we managed this successfully, probably because we both possessed a good sense of the ludicrous, so it was fun. Then for my birthday my father gave me £30 towards my A licence. Where to spend it was settled quite simply by the family camping for the summer holiday in Henry's barn in north Devon, on the hill above Barnstaple aerodrome (now part of Chivenor). I would learn at the flying club there.

Barnstaple was a small grass field bounded on two sides

by the estuary. It had a little hangar and clubhouse, a small number of Moths in blue and silver paint and two instructors: J.W. Nash and R.J. Boyd. I flew only on one Moth, G-AAIM, and almost entirely with Nash. I do not think I was a very good pupil because being in the air was still like a blissful hazy dream, but I do not think either that the club was in any hurry to expedite my training as there were few other pupils and they needed the money. So they invented expeditions such as flying to Lundy Island and landing in a little field with sheep – called Lundy aerodrome; probably not realising that I was running out of money faster than they were. Each day I would freewheel down the hill on my motorbike to save petrol, but returning home was a different matter as the petrol tank was circular with the fuel outlet forward of centre. Unless the tank was well filled when going uphill the petrol sank downhill of the orifice and the engine died. This happened quite often.

Eventually, on 5 September 1934 I went solo. Nash got out and did up the front cockpit straps while my happy dreamy state took on a disconcerting blankness. I think he must have told me what to do but I do not think I heard him. As he backed off from the Moth and gave me a wave I opened the throttle mechanically to turn and taxi close to the downwind hedge, the other side of which was the water. Then I opened up fully, the Moth bounding into the air as though it, at least, was pleased with itself. All too soon I was gliding down again over the shiny mud and grey water of the estuary, over the same hedge, and was back on the ground and taxiing in. I was so happy that I was speechless – and Nash was probably glad to get his Moth back intact.

But this was the end of the summer holiday and although I had flown solo there had not been time, mainly as a result of days of wind and rain, to get my licence. So back home I joined Brooklands, which the BSA put within reach, and squeezed a little more subsidy from my father. I soon found that flying at this mecca of racing tradition was somewhat different from easy-going Devon. The club members all seemed to know each other by their Christian names, with life one long party. They were kind to me but my single-minded approach to flying, and my youth, did not easily

PORTSMOUTH DAWN PATROL MAY 30 1937

LEAVING BROOKLANDS ON A CUP OF TEA WE FLEW BY DEVIOUS (VERY) ROUTES TO PETERSFIELD POND WHERE MOST OF THE ATTACKERS HAD SECRETLY CONSPIRED TO MEET

AND AT 3.40 (BY KEN'S VEREY LIGHT) WE FORMATED FOR PORTSMOUTH 12 AIRCRAFT. IN ARROW HEAD FORMATION, WITH 2 PUSSES AND A LEOPARD ABOVE AND BEHIND

AS WE APPROACHED PORTSMOUTH THE THREE UP BEHIND DIVED AWAY AND ATTACKED THE DEFENDING AEROPLANES. WORRYING THEM AWAY FROM THE MAIN FORMATION. (ON THE CONDITION THAT WE SUBSCRIBED TO THEIR BACON AND EGGS.) AND ONLY THE TWO OUTSIDE MACHINES WERE SPOTTED (UNFORTUNATELY JIMMY . ME)

← INSTRUCTOR PASSING OUT

REMARKS FOR POSTERITY
"AND DO YOU KNOW I COULDN'T SEE A SINGLE AEROPLANE, NOT ONE, UNTIL I FOUND I WAS FLYING UNDER THE HOOD"

THEN WE ALL LANDED FROM DIFFERENT DIRECTIONS AT THE SAME TIME AND OVER AN EXCELLENTLY SERVED AND BETTER APPRECIATED BREAKFAST, COMPARED NOTES WITH THE OTHER WORM CATCHERS.

adapt itself to the sophisticated socialising of this famous place. But the flying was marvellous, and I was looked after to begin with by 'Mack' and Duncan Davis on Moths G-ABBW and WN, qualifying for my Royal Aero Club Aviator's Certificate (No. 12413) on 8 October 1934; which then included gliding without engine from 2,000 ft and hoping to make it into the landing area. My Ministry 'A' licence (No. 7226) was issued on the strength of the Royal Aero Club Certificate.

During the whole of 1935 I could afford only about seven hours' flying but it was two or three circuits regularly twice a month. I managed to sell some advertising artwork to *Flight* magazine and Fairey Aviation which helped, but it was a lean year. 1936 was better as, with some thirty hours' total flying, I was accepted as experienced enough to fly new aeroplanes brought to Brooklands by salesmen, and also to go on 'breakfast patrols' on club aircraft. Two of us would share the cost, one flying to whatever aerodrome was being 'attacked' and the other back home. I also managed to do a little flying on Aeronca C-3s at Hanworth; so that year I logged a new Tiger Moth G-ADEL, a BA Swallow, G-ADOB, BAC Drones, EAN and EJH with 6 hp Douglas engines, a Moth major, Taylor Cub A-AEIK and Avian G-ABVG. Then C.G. Grey bought some of my cartoons for *Aeroplane* magazine, I took my first passenger and looped him, earned a bit more money as assistant in a car showroom selling 30/98 Vauxhalls, and started to learn slow rolls with Max Findlay. On the first one, as we became inverted, Max came quite slowly out of the front cockpit until his head rested on the underneath of the fuel tank. A gurgling sound came through the gosport tubes which I could not understand, nor did I know what to do; so I did nothing. The roll continued and, years later it seemed, the gyrating horizon returned to somewhere underneath and Max sank back into the cockpit; followed by a few minutes' silence while he did up his harness.

There is no doubt that flying in the 1930s possessed a certain individuality. On one dawn patrol into Portsmouth from Brooklands we 'attacked' *en masse* having arranged to formate over the town of Petersfield at eight o'clock on a

Growing up

Sunday morning, with no thought of the noise made by a dozen aeroplanes at 800 ft. Going into land at Portsmouth Jimmy Gunn, who had this first half of the flight, undershot and hit the main Portsmouth road before bouncing over the fence to arrive neatly just inside it. A few weeks later I went with Ivor List in a Puss Moth to his old school at Haileybury where we landed on the playing field, which turned out to be a little smaller than he had remembered when he had been a pupil there. Taking off we failed to clear the tops of the oak trees and needed to land in a far corner of Brooklands to remove the twigs and leaves before showing our faces in the clubhouse.

One day I was in the front seat of a Moth when the pilot lost control in low cloud. With nothing to do except look out I immediately yelled as the trees on the top of Box Hill shot out of the murk. We screamed over them as we came more or less level. It was the nearest I had been to a 'close call' and it heightened the pleasure of our lunch in the warm sun.

Brooklands was a very relaxed place. There was no ban on drinking and flying; it was reckoned that as every pilot would probably sooner or later find himself flying after a party he might as well learn to do it properly. So the bar opened every lunchtime with Pimms as the favoured tipple. It was up to you whether you drank or not – and my first real lesson in taking decisions for myself.

2 · The good west wind

By the middle of 1937 my log-book held 50 hours of flying, including some as a passenger with anyone who would let me hold the stick for even a few minutes. Then a body calling itself, I think, the Anglo-German Fellowship, circularised information about a gliding course at Dunstable. It was cheap, though most aeroplane pilots regarded gliding as mere aerial tobogganing and not real flying at all. But any chance to get into the air needed to be taken, so I went. This time my kind mother, who supported all my strange activities, helped me with money, balanced against the food I would not be eating at home for two weeks.

I found my way to Dunstable by bus, walking the last few miles out to the Downs, looking with fascination at the slow silent gliders floating over the bare hillside, their wings transparent against the bright summer sky. This did not look like derisory tobogganing to me. I soon discovered that gliding was not only flying but something you became a part of. There were no mechanics to put the aircraft away at night; with other members you had to cunningly fit into the small hangar more gliders than seemed possible. The instructors were ordinary club members, not paid professionals; and when gliders were broken we all carried the bits to the workshop and helped mend them. Every day I was out on the hill in the soft summer wind flying or launching other pilots, running hard, stretching out the rubber bungie catapult, ducking as the whistling glider shot overhead. Somehow it all had a great creative feel; it was a new world and I felt I belonged, helping to build it.

There were twenty-five on the course – sixteen young Germans and nine British. I was the only girl. We started by flying Daglings, very basic gliders with no cockpit and no springing to absorb bad landings – a real encouragement to learn! The first flights lasted only a few seconds, barely time to breathe after the sharp acceleration of the launch before the ground rushed up and you hit it – a normal landing was one which did not actually break the glider. We flew in turn, launching each other with the instructor, Tim Hervey, shouting the commands which soon became part of our existence: 'Walk – Run – Let go.' This last meant that the person lying on his tummy in the grass hanging on to the glider's tail let go, not those pulling the bungie! Off went the Dagling on its ten-second flight, with us in pursuit to get it back up the hillside so that our turns would come round again sooner.

After a few days we were allowed to take off from higher up the slope, and then if we had not broken anything, right from the top; and although getting the glider back up the hill now required more energy it did not happen so often. There was also a rudimentary winch operated from the hilltop which did the actual pulling of the glider so that it was only necessary to walk up the face of the Downs holding the wing tip. That was the theory; in practice lack of communication ensured that the winch started before you, out of sight at the bottom, had properly hooked on your glider, or it pulled so fast that mountain goat leaps became essential to keep up with it, or it left you stopped and forgotten in mid-ascent – usually because everyone at the top had gone off to fetch a glider which had landed on the golf course; or – final indignity – the winch did not stop when you arrived and you were left with the exciting task of getting it unhooked before it rounded the wheel at the top and left you hanging on to a run-away aircraft. Much depended on whether someone on the hill wanted your glider; it was all much better if they did. Flying an open Dagling from the top of Dunstable was exhilarating and to begin with a little frightening. After the wild acceleration of the launch you had somehow to slow down to the right speed; but not too much and stall – without an airspeed indicator. But you tried not to go too fast either, as this got you to the ground more quickly and lost precious seconds of airborne time.

As the days passed, warm and sunny, we were advanced to less basic gliders. One RAF pilot, Widdle Shaw, quickly moved to the Falke, a stocky floater with no upward view, while I was promoted to the 'Boot'. This was just a light cockpit fairing clipped on to a Dagling, to improve the performance enough to allow it to soar over the hill in the lift from a fresh west wind. Proficiency as a glider pilot was based on certificates: A for a flight of 30 seconds, B for two 45-second flights plus one of a minute, and C for 5 minutes' soaring. My first flight in the 'Boot' for my C produced no 5 minutes as I flew too far out from the ridge, not wishing to cartwheel into it as I had seen others do, and so sank quickly down to the landing field. Getting back up to the top again took a while but the westerly breeze was still blowing, and with fierce instructions from Tim Hervey to *hug* the hill I was shot into the air and turned quickly along the ridge. This time it worked, the grass and the little bushes rushed past just underneath and I just achieved the magic 5 minutes. I got my A on 16 August 1937, my B on the 17th and my C on the 18th.

It was not just our own progress which was exciting. Gliding was just beginning to flourish in Britain as the delights of cross-country soaring on the thermals under cumulus clouds were discovered. Almost every weekend exploratory flights were made from Dunstable and sometimes records broken. It was pioneering flying and even as a beginner I was part of it, helping to launch the explorers – Philip Wills, Kit Nicholson, Hugh Bergel, Sebert Humphries – and sometimes going off in an open touring car to retrieve one of them from some strange field. I loved every minute of it; the unexpected, the new, the belonging.

After the course I returned to Dunstable every weekend. My parents had sold our house at Bickley and lived, at that

The good west wind

time, on Ham Common in Surrey, and each Saturday I set off at 6.0 a.m. to walk the 3 miles to Richmond, go on the Underground to London, and walk to the Bergels' house for a lift in their Riley to the Club; reversing the process on Sunday evening. During the autumn I progressed to the Falke and then to the Grunau Baby, which was easier to fly, and during December I tried for the 5-hour Duration flight for my Silver C. This was the most advanced badge at the time – and for which, in due course, I would need to fly 50 km distance and climb 1000 m to complete. It was a day of snow flurries which helped to pass the time because of the need to be at the end of the ridge which was in the clear while the snow was blotting out the other end. Even so, for the first hour while my feet became steadily chilled I was convinced that time had ceased moving. I finally landed after 5 hours and 3 minutes cross wind, cracked the landing skid, and spent the next day mending it.

Sometime during that winter I became an instructor. With no qualifications needed it was something that just happened to almost any member who turned up regularly, and it began by my being told to look after the top bungie launch point and offer any advice that seemed helpful such as 'Don't hit the retrieve car that's run out of petrol in the middle of the landing field.' But once started that is the way it went on. Newer members assumed you to be an instructor, but if all the other instructors were flying someone had to give even minor useful hints. So close to my own learning mistakes the knowledge I passed on was at least fresh and practical, although undoubtedly lamentably weak in theory. I discovered that I enjoyed instructing. With whole days up on the hill in the fresh wind there was time to both fly and teach, and it was fun thinking about how to explain the best way of flying a glider from the top of the hill to the bottom without creating a disaster on the way.

Whenever possible I still flew aeroplanes, and one day Widdle Shaw suggested we went to the 1938 Empire Exhibition at Glasgow, but I could spare only Monday and Tuesday, and Widdle only Tuesday and Wednesday, so it just had to be fitted into one day. I left Richmond very early by bus and train for London, and by badly misjudging the weekday speed and frequency of those vehicles arrived half an hour too soon. We left there at 8.01 by car, eating our breakfast of fruit cake and apples as we drove up the Great North Road to Hatfield. The hired Hornet Moth's engine was being run up as we arrived and at 9.13 we were in the air, soon flying over partly built military aerodromes which we marked on our map. Near Leicester, in a rain shower, I had a good look at the ploughed field in which I had landed a glider; it looked much larger and easier than when I had first seen it, from 600 ft in a 10 ft per second downdraught! As the wind was north-west, rough and fairly strong our ground speed was only 60–70 mph under a sky full of big cumulus throwing purple shadows like jigsaw pieces over the ground. We did not really mind that the flying would take longer than anticipated, but because this Hornet was fitted only with single controls and we both wanted to fly we had to change seats. This was done by the passenger, myself to begin with, crawling backwards into the baggage shelf, while the pilot clambered sideways over the central arm into my seat; I then slid forward into the left seat. There was cloud on the Pennines so we flew round them still making only 70 mph over the ground – we had hoped to reach Renfrew by 12.30 but were obviously going to be late. After some dirty rain near Manchester we came to the coast south of Morecambe Bay into brilliant sunshine. Little white cumulus were sailing across a clean sky above a deep green sea, and as the tide was out we flew low over miles of rippled and shining sand, very pleased that we did not have to walk all that way for a swim. We changed seats again just before crossing the coast at Furness, with Widdle becoming stuck in the baggage shelf while I began to worry about petrol; we had been in the air for nearly 4 hours and the gauge could not be persuaded to show more than 7 gallons. Barring our way inland the dark Lakeland mountains wore majestic cumulus crowns, but ahead towards Carlisle there were foothills facing into the strong sea breeze, so we flew straight along the face to collect an extra 1000 ft free from the updraught. The petrol gauge subsided to just under 5 gallons but the hills were also subsiding so we went low over them and made Carlisle aerodrome with just 4 gallons. We thought we might have to circle the town to encourage someone to come out and give us petrol as it was lunchtime and a weekday, but there were people there in plenty, so we had a cup of coffee in the clubhouse, got the weather, and sent a telegram to Renfrew for a taxi at 3 p.m.

We took off again at 1.45 and flew a compass course over the Scottish Lowlands in perfect visibility. We changed over again and I was flying, while Widdle put a name to every peak in sight, including the island of Ailsa Craig 50 miles away. Our first sign of Glasgow was the Empire Tower sticking up in the sunshine. We reached Renfrew at 3.01 p.m., landed, leapt into the waiting taxi, and by 3.20 were through the Exhibition turnstiles.

The first thing we did was to run into a friend, who complained that he had spent most of the previous day travelling and would have to take another day to get home again. We gave him a superior smile and asked him the whereabouts of the new Hawker Hurricane exhibit we had heard about; but he had not seen it and we never found it. Quickly we did an ordered round of the pavilions, including *The Times* building with its wonderful pictures of clouds, gliders, and infra-red air photographs, then bought some Edinburgh rock and an Exhibition programme to prove we had got there. We now had half an hour left so made a dash for the fun fair. We tottered off the dodgems and went on things which went round and round and up and down very fast – which made us feel sick. Then we tried a little race car track and Widdle had a head-on crash with an irresistible object, pushing his bottom teeth through his chin; so we went on the wall of death train for a rest.

We got a taxi back to Renfrew at 6.10 p.m. and within ten minutes were in the air heading home downwind in a clear sky. I was flying, down to 200 ft over the Solway Firth and up again over the Lakeland foothills, whose mountains still wore a halo of cloud. Far to the south-east clouds lay like pink hills. As we reached the Pennines, topped by big cumulus, we climbed and at 4000 ft cleared their crests. They broke beneath us, but there were more ahead, higher cumulus sprouting false cirrus; so we went up to 6000 ft, changed places again and played among them crashing into dazzling white mountains, diving in dark valleys, and searching for our haloed shadow on towering walls of snow. Suddenly we decided to visit the Derby and Lancs Gliding Club ridge at Camphill, so we spiralled into the dullness below to spot Bradwell Edge with four gliders floating gracefully and seemingly motionless 1000 ft above the hill. We shouted with envy and roared low over the people at the launch point. They waved as we closed the throttle to sail along the ridge face, pretending to be a glider like those four now high above us. But we failed to soar and, when down to hilltop level, departed.

With the good wind behind us we were doing 130 mph over the ground, and soon Sywell appeared and later the new Luton Airport. We flew on without talking, content in the cabin of our little world, while outside the bright June day faded. High overhead the pale sky was swept by fine threads of golden cirrus while on the western horizon the low sun glowed through a rift in a dark and watery bank of stratus.

Ahead on the misting earth we saw the flash of the Hatfield airfield beacon, and a few moments later slipped in to land beside the hangars. It had been a great day and neither of us felt tired. Our flying time was 8 hours 10 minutes with $2\frac{1}{4}$ hours actually in the Exhibition and $1\frac{1}{4}$ hours for refuelling, getting forecasts, and the taxi rides in and out of Glasgow. We patted the friendly Hornet goodbye and went to get our first meal since breakfast.

Back at Brooklands the College of Aeronautical Engineering had built a Grunau Baby glider as a student exercise and they sold it to me for £50. Its test flight was a brief bungie hop beside the hangars. A glider, of course, needs a trailer so I designed one – just like all the others – and found a jobbing carpenter to make it for £20, using an old Morris car back axle and wheels. I told him to use casein glue which had moderate water-resistant properties but he used fish glue, which had none.

I was very happy with my varnished Grunau, but during an expedition to explore the North Downs that spring, bungie launching from the top of Colley Hill near Reigate, it tried to bite me. Without warning it spun for no reason. I knew I had enough speed although the Grunau, like every other glider I had flown, was without an airspeed indicator. I recovered quickly and after landing we all had a careful look at the wings – to discover that the twist in each was different. The College was happy to rebuild the defective wing but I was not happy to lose my glider for several weeks as we had more expeditions planned. With slow, light gliders, such as my Grunau, we could launch from almost any good-looking hill, and this summer we wanted to spend our holiday wandering among the chalk downs of Salisbury Plain which we reckoned would produce plenty of thermals.

We finally set off, after the usual chaos of getting organised, on 22 July for Huish, near Marlborough, with Widdle's car, my Ford 10 and Grunau, and a winch. This was an old Chevrolet with a drum mounted on the transmission shaft, and everything behind that cut off and replaced with a tow bar. It also carried an emergency cable-cutting axe. The next day Geoffrey Stephenson (Steve) and Donald Greig arrived with their Grey Kite and Jack Dewsbury with the Rhönsperber. We first flew from Roundway Down in such a light west wind that Steve sank below the top, so instead of going all the way to the bottom he made a downwind landing towards the hill face. When the wind eventually strengthened we decided to launch Greig (he was never known as Donald) from this rather rough place to save carrying the Kite over two fences. A very small girl was found to hold the wingtip, I held back on the tail, and the other two dashed down the slope stretching the launching bungie. Greig shouted to let go, I released the tailskid, there was a long pause, and the rubber rope shot back up the hill with the launching crew helpless on the end of it. The Kite had not moved. We tried again, three of us pulling and no one holding back, the little girl still at the wingtip with an expressionless face. This time the Kite crept two yards and I fell into a hole; so without consulting the pilot we fetched the Ford V8, attached the bungie to the bumper and drove flat out for the edge of the hill. The Kite shot into the air with such violence that the released bungie dented the back of the Ford as it stopped two feet from the brink. But the

The good west wind

glider was in the air. The next day gave us weather good enough for a cross-country flight. I was launched and departed first only to land in a ploughed field 12 miles away, but Jack Dewsbury in the Rhönsperber reached the coast at Faversham, 130 miles.

On its next flight my Grunau disappeared while being soared by a friend from Huish Hill, so Bill Murray, who had just arrived to fly it, and I went off on a search. After asking several road menders and doing much trailer reversing in muddy lanes we discovered it 5 miles along the ridge lacking its elevator and nose, having been landed on a large and solid fence. In silence we surveyed the remains and packed them into the trailer.

Arriving home we made frantic telephone calls to every glider repairer in the country, but it was a bank holiday weekend and our only hope was Yorkshire, so at 10 p.m. Bill Murray and I left in his MG Midget for Kirbymoorside. We burst a trailer tyre at 1.30 a.m. and another just as a very cold dawn was breaking, but arrived at Slingsby's just as the Works opened. After a welcome bath and breakfast at Sling's house we surveyed the damaged Grunau and set off up to Sutton Bank to collect a Kite which Fred, in his usual generous way, had lent us so that we would not miss any flying. There we changed the trailer fittings to suit the different glider, packed it in and returned to Huish the same night, arriving at five in the morning: 540 draughty miles in 31 hours. Looking back on these years our single-minded energy seems unbelievable. We never stopped for an instant even though our future for us still stretched into infinity.

Next day the wind blew from the east so we followed it round the hills to Inkpen. The weather was hot but the lift feeble so to lighten the gliders we left cockpit covers, coats and parachutes on the ground, and Bill was flying in this state of aeronautical nakedness when a huge thunderstorm brewed around us. Quickly we launched Steve into the black sky and gusty wind – now blowing down the hill – but although the Kite was twice shot up 20 or 30 ft with great violence he was unable to connect with any lift. Both gliders landed in near darkness as the rain began.

Our last day was spent back on the top of Roundway in perfect weather but for a lack of wind. Bill got a thermal off his winch launch and departed, Steve did likewise although the cable broke at 300 ft, and I had to go home for the day. Greig now found himself 9 miles from our base at Huish with two empty trailers, a winch, 2000 ft of cable straggled across the grass, and two cars, but after a good think he got everything back to base *and* had retrieved Steve from a 33-mile cross-country by 11 p.m. I arrived back at Huish at midnight to find that Bill had soared almost to Dunstable, 55 miles away, in the Kite, so next morning I drove there to collect him. At midnight (again) Bill and I left for Kirbymoorside to return the Kite and collect the Grunau. It rained and thundered all the way, the MG's petrol pump packed up, and the fish-glued trailer came apart at the seams – in spite of which I managed to get some muchneeded sleep while Bill was driving. We arrived at the factory, again just as the doors were opened, but the Grunau was not yet ready. As the trailer was also now in need of repair we had to forgo our plans to drive to Scotland the same day, to try out the great 1200-ft-high Bishop Hill in Fife. But the following morning with the Grunau safe in a trailer girdled with metal hoops we set off, and after two punctures reached Bill's parental home at Stirling just in time for dinner – only to find that the glider fuselage had come adrift inside the trailer and had a hole in its plywood side. By midnight we had the repair finished, but I think Bill's parents thought we were mad.

Next morning, after a good breakfast, we towed the trailer the 30 miles to Bishop Hill; examining the steep ridge with a critical eye as we circumnavigated it to find a way up the gentler slope at the back. Eventually after several excursions along roads which soon disintegrated into sheep tracks we discovered the right one to a farm $1\frac{1}{2}$ miles from the crest, where the Scottish Gliding Union kept their equipment. They were just starting to explore this hill and we wondered why they had so far made only one soaring flight from it. We soon found out. Leaving the trailer at the farm we went on in the MG to have a look over the edge, but the road soon ceased to be even a cart track. It crossed holes in stone walls and dropped down deep ditches. We bogged the MG, fetched the gliding club's old bull-nosed Morris, bogged that and finally walked to the top. The view was as beautiful as the hill, and far below in the valley Loch Leven glittered in the afternoon sun.

The following Saturday, with Andrew Thorburn and some of his club members, we set about trying to get the trailer to the top, but even with their help it took $5\frac{1}{2}$ hours to travel the $1\frac{1}{2}$ miles. We tried towing the trailer with two cars, we dug a road, and we burnt out the old Morris's clutch. Wheelspin was our worst problem so we roped the wheels, which promptly dug themselves in worse than ever. Eventually we gave up and Andrew went away, to come back a little later with a horse. It galloped everything to the top in 20 minutes, but as it was now too late to fly we left the trailer in a disused coalmine. Next morning we rigged the Grunau in a lovely west wind, and I was just about to be launched when it backed to the south-east and stayed there, blowing down the hill. We de-rigged.

On our next attempt the wind was again west, blowing about 40 mph at the top, but as we finished rigging the clouds swooped down all around and the rain came; we de-rigged. The following morning the wind was even stronger, blowing at well over 50 mph at the edge; but the next day was perfection. The wind was about 40 mph and I was launched by someone under each wing carrying the glider to the brink and just letting it fly out of their hands.

I had never before flown from such a big hill and the freedom it gave to wander out over the valley and its silver loch was a revelation. After some two hours I landed close to the edge to give someone else a turn, but no one came. I did not dare get out as without my weight the light Grunau would have blown over, so I sat with my wings rocking in the gusts. After a while this became boring but still no one arrived. They, too, had got tired of waiting and had gone off to find some shelter from the wind and with any luck a cup of tea. Suddenly a slightly bigger gust hit the glider and while holding the nose down it moved a foot or so forward, right to the edge where the ground fell steeply away. Well, if no one else wanted to fly, I did. So I waited for the next gust, let it lift the Grunau clear of the grass and eased the nose down just enough to accelerate but not quite enough to return to earth. We teetered for a second over the edge then hit the huge upblast of air rising up the face of this big hill, and in it had another glorious hour's flying.

By now Widdle Shaw with a Falcon III two-seater was on his way from Yorkshire, and profiting from our experience brought it all the way up with a two-horse team, making the journey in 2 hours. In the afternoon the wind slackened to 25 mph but we were still able to launch by hand. The lift was absolutely smooth and I could stall turn and loop without losing any height at 2500 ft above the valley.

On our last day of all, Sunday, we flew again but the wind was too light to soar and there were heavy showers. The flat top of Bishop Hill was large but there were only about three practical landing places, each about the size of a tennis court, and these were for the most part covered in small boys. A winch had now been persuaded to reach the top, but on one launch the cable went slack when I was at 30 ft over the coalmine; thinking the wire had broken I released, only to find that the combination of low sun, breath, and rain on the cockpit cover had reduced visibility to nil and I could see no ground at all. My poor Grunau bounced heavily on the springy turf of a little ridge and ran sideways down a twelve-foot bank into the edge of a bog – but nothing broke. Shortly afterwards the wind died on Widdle in the Falcon III and he sank down into the valley. As we were leaving for London next morning I was winched over the edge to fly down to join him. Then it poured with rain.

After these expeditions I started thinking seriously about starting a gliding club myself as I had lots of ideas and wanted to try them. The flying we had done from Colley Hill had shown that the North Downs were good for gliding and that a club there for people living south of London would be popular; as Dunstable was for the north. The London Club had some thoughts about starting a southern branch but nothing came of it; in the meantime the owner of the Redhill Flying Club, Graham Douglas, expressed interest and after finding how impatient I was to get a club going he agreed to put up £300. The proposed flying field would be at Buckland in the valley below Colley Hill which was owned by a Mr Sanders. He became a partner in the venture and we were in business. We had an inaugural meeting to discover local interest – there was a lot – and then I bought equipment: from Fred Slingsby, one Dagling, one 'Boot' Fairing, and one Kirby Kadet secondary glider; a 50 x 20 ft wooden hangar, delivered flat; a cut-down car with a drum mounted on the transmission shaft for a winch; 3000 ft of steel cable; and a fifth-hand bull-nose Morris for field retrieving. There was just enough change from the £300 to have leaflets and log-books printed. In October 1938 the new Surrey Gliding Club started flying from Buckland with me as chief – and only – instructor. I was just twenty-one.

New members came flocking and while teaching them to fly I had to teach myself the techniques of training with a winch from the valley instead of by bungie from the hilltop; though both methods had the same disadvantages of learning solo – of making mistakes when very close to the hard ground. So it was essential to give the new pilot plenty of ground slides to become familiar with the controls – though not the rough ride over the bumpy field provided by the unsprung seat! When he could keep the glider running straight the winch could be driven a little faster so that the beginner would fly at about 5–10 ft. This low hop stage caused the instructor driving the winch to quickly acquire the skills of an expert fisherman. It also indicated clearly that more instructors would be needed soon if my nerves were not to be too sorely tried. One of the first was a tall seventeen-year-old, Brian Powell, who had started flying at Dunstable. His first-ever flight in anything was from the top of that hill when still sixteen; the instructor sent him off without asking him what he had flown before – or even if he had! Unaware and undaunted he luckily made it to the bottom with a dull thud but no injury. He became a good instructor, well able to remember what it was like to be a pupil!

We had, in fact, surprisingly few accidents. One beginner got a bruised back, both Graham and Brian subsided into the trees on the top of Colley Hill, and one member overshot and flew straight into the wall of our club hut, about half-way up; only the hut was damaged. The club spirit was wonderful. Members built the hangar, and resited the hut, which the club had been given, by carrying it bodily down the steep face of the Downs, while others constructed signalling bats and all the other bits and pieces that seemed to be needed. One member who made much possible was Dukinfield Jones. He just appeared one day, a slight grey-haired man with glasses, saying that he would like to join the club and could become our honorary secretary if we needed one. He also wanted to fly again having been a Vickers Gun Bus test pilot in the First World War. Nothing could have been better and I relinquished the paperwork with joy. He not only did it much better, but kept detailed records of club flying and members' progress. When he died forty years later, aged eighty-eight, he left £1,000 to the Surrey Gliding Club at which he had been so happy.

It was now summer 1939. My little club had grown to 100 members, many of whom had gained their A and B certificates. We flew and flew and the summer weather was kind to us. Midweek I put in hours on the Moth Minor G-AFOX at Redhill, the delightful little Chilton G-AFGI and a gentle elderly Moth, G-AAAC; and in the middle of all this Graham Douglas and I were to be married. My parents invited 150 people to an engagement party for me; but this was the day that Geoffrey Stephenson made the first ever soaring flight across the English Channel, from Dunstable. His partner, Greig, did not have a passport, so Brian Powell and I left the party in mid-celebration to retrieve him. Wearing the blinkers of youth it did not even occur to me that this was an ungrateful thing to do, particularly to my warm-hearted Mother. My only consolation is that I am sure she understood.

3 · Downhill, uphill

It had not been only flying in those marvellous pre-war years, freed from the conventions and constraints of childhood. In 1936 I had met a young German at a party, called Jochen Benemann, who told me of a skiing camp he was organising at the end of the year. Would I come? Since the weather was not usually kind to flying in early January it seemed a good opportunity: I had never tried skis before, and it would cost only £16 for two weeks total, including ski repair – I broke two. I met the rest of Benemann's party at Victoria Station and we travelled by boat and wooden-seated train to the Rossfeld Ski Hut above Berchtesgaden, in the heart of Hitler country; about eighteen young Germans and seven English. The skiing was primitive, climbing uphill as there were no ski lifts followed by exhilarating and uncontrolled descents. It was a wonderful two weeks, falling about in deep soft snow all day, and eating, drinking and talking in the warm hut in the evenings; it was essential to get really warm before going to bed as the bunkrooms were not heated. But some of the talking was. What we had not realised was that the Germans were trying to woo us so that we would not want to fight them in the event of a war. There was much else in our innocence; we were totally ignorant about politics – living for the day was fun enough. The German contingent kept on at us to have an evening of serious discussion on Anglo-German relations, but this was not what we had come skiing for, and when we said so they sulkily drank their beer while we hilariously quaffed ours.

It was Benemann who had told me about the 1937 gliding camp at Dunstable, and, nothing if not persistent, he now arranged another ski camp for the following January 1938, rounding up some of the English glider pilots.

Again we went to the Rossfeld, with the Germans stepping up their propaganda by walking us there past Hitler's eyrie. They tried to arrange for us to meet him but failed, though they succeeded in getting us to a lunch given by Baldur von Shirach, who made a speech we could not understand. Benemann again tried the political discussion ploy which he had promised us he would not do, so this time we all sulked over our beer for a short time to show our displeasure. Of course we thought that there might be a war, the newspapers too often wrote about European upheavals for even our scanty reading of them to miss. But in the days before television it was hard for the young to visualise a war in which they might be involved, and parents' tales of 'The Great' war possessed the remoteness of history. If we thought about war it was in terms of something exciting, but mostly we did not think. Even the uniforms all over Germany did not alarm us, the goose-stepping had a faintly ridiculous quality – and the lucky Hitler Youth had their gliding free. To us, easy-going English, there was even a certain attractiveness about German efficiency; things worked. Coming out in the train my passport had slid out of my pocket in the dining-car, which had then been detached and disappeared to some other part of the Fatherland. At Munich station James Hayes, who spoke German, and I reported this to the Station Master. With no fuss I was given a document covered all over with impressive rubber stamps to use as a passport while they tried to find mine. For several days I heard nothing, and reckoned it had gone for ever. Then one morning a diminutive Hitler Youth arrived at the hut, having trudged 1000 ft up through the deep snow complete with bare blue knees, and personally handed me my passport. I was amazed.

Shortly before we returned home Benemann tried his wooing tactics again, this time producing Rudolf Hess to talk to us. Wisely, perhaps, Hess said nothing political, but drank beer with us, talking about our skiing and gliding. He was like no other German I had met, having those slightly hooded, far-away eyes of a dreamer; while all the other officials we met had rather blank, expressionless eyes. After a few hours he left with a friendly farewell to ski down to Berchtesgaden before dark. We watched him go until he was a speck in the distance, as he seemed to be just as good as we were at falling over and making great craters in the snow. Benemann thought we were tiresome but, presumably as instructed, he never gave up and arranged yet another ski holiday for the first weeks of 1939.

Again most of us were glider pilots and this time we went to a ski hut above Kempten, as Berchtesgaden was by now probably somewhat too sensitive. Our journey out was as usual by train but this time there were no wooden seats for us; they were all occupied by uniformed Germans. They occupied most of the corridors as well and we had to stand all the way with barely room to breathe. This had the disadvantage of preventing access to either the train restaurant car or station buffets and by half way across Germany we were starving. One of our Dunstable gang, Alec Barnard, who liked nothing better than a well-ordered meal in the best restaurant, was feeling particularly deprived, so we deputed him to use his size and weight to work himself to the end of the corridor before we reached the next main station, get out and quickly buy all the food and drink he could – Alec always had plenty of money – and pass it to us through a corridor window. All he had then to do was to jump back on the train. Everything went according to plan until Alec, having passed us huge bottles of beer and satisfying bread and cheese ran to the carriage door; but it had jammed shut. In a panic he returned to the window imploring us not to leave him behind which we thought highly funny. There was only one way to retrieve him and that was through the window. As the train started off we pulled and Alec scrabbled with his feet up the side of the coach and in he came – head down. It took a long time in the press of people to get him the right way up, by which time most of the beer had been consumed to provide energy.

The new ski hut was good and so was the weather and the snow, but the Germans' humour seemed to be deserting them. Ours seemed to surface whenever they saluted each other while skiing and fell over in their bright uniforms. When we fell over they responded with derisory expressions. We thought they were becoming pompous and reckoned we would not bother to come back next year. They thought we were politically stupid, which was perfectly true.

That summer of 1939 had a breathless quality. More and more members were joining the gliding club, gaining more certificates, and we were able to buy more and better gliders for them to fly. Memory says the sun shone every day, and everyone was happy with their flying. Being young we were not worried by the now imminent war. We talked about it often, handing out strong opinions founded on almost no information, and we knew in our hearts that it was coming; but not when. Nor did we have the remotest idea of how it could or would alter our lives; but we possessed no foreboding, no dread. My imagination at least did not stretch to overwhelming disasters. Anyway, it was not us that wanted war, it was other people. So we continued to fly our gliders, happy with every day as it came.

At the beginning of September I took my little Ford 10 tourer to Torcross in Devon for a holiday with my parents, driving overnight on the narrow twisty roads for the eight-hour journey. This time I was stopped at frequent intervals by police or people with armbands, waving torches, and looking for Germans. I arrived at my aunt's house in time for breakfast on one of those perfect mornings of late summer, the sun warm but softly hazy and still low enough to glitter a path across the sea. For two days I swam with my mother or lay in the sun listening to the rhythmic breathing of wavelets on the shingle, while my father fished happily for mackerel from the beach; they came in shoals chased shorewards by leaping porpoises. Then it was 3 September. We heard Chamberlain's announcement of war and the holiday was finished. After lunch that day I set off back home taking my father who was still in the Army reserve, running the gauntlet of more posses of self-important German seekers. I was met by Graham, already in RAFVR uniform, and that night we started defending Redhill aerodrome. In retrospect it was quite ludicrous. In shifts of two with one service revolver and a tin hat between us we walked round the perimeter in the dark, through the woods and in and out of the hangars, stopping with held breath at every strange noise. Any Germans, in uniform or as nun-clad parachutists, would have been defeated only if they had died laughing. But what else was possible? Although Redhill now had RAF Fairey Battles with which to check out escaped Polish pilots, still flattened by their last few weeks of disaster, there was no military security organisation, so it was either us, two at a time, or nothing.

Next day I went to the gliding club to find Dukinfield Jones in charge of packing everything away in the hangar, helped by long-faced members, not knowing if they would ever fly again. I looked at my little varnished Grunau in its trailer. It now had two instruments: a clock from a First World War Sopwith Camel that a friend had given me and an ex-RAF airspeed indicator which read 200 mph when the speed was actually about 40. I left them there and locked the trailer, hoping I would fly my Grunau again, but at the same time knowing that I would not.

A few days later the police came to collect my Austrian cook. She had been produced by my mother-in-law, probably to ensure that her son was properly fed, and was taken off in tears to an internment camp. As she had only just escaped as a Jewish émigrée from a dictatorship, her despair was understandable. She left me, who had never cooked anything more ambitious than sausages on a barbecue bonfire, with a raw chicken. Having to learn to cook was not my idea of war at all, but with civil flying forbidden time dragged. Looking around for something more interesting I was given a shiny black tin hat and the job of checking volunteers as ambulance drivers! Later, in the spring of 1940, the elementary flying school at Redhill was posted to Carlisle with Graham as its Commanding Officer, and our home was requisitioned. My kind mother came to help with the sudden packing, and for the rest of the war looked after our belongings and the dog. Time dragged also at Carlisle as there was little for me to do. I went to art school to study lithography, and I was 'arrested' on Crossfell mountain by an armed soldier for 'heliographing to the enemy'. He reported me to the CO at Carlisle – Graham – who had a problem keeping a straight face because he knew I had been there to photograph the lenticular wave clouds produced in an east wind by the mountain itself. Slowly the summer of 1940 passed, nourished by rumours of non-existent attacks, and isolated moments of interest, as when a Fairey Battle made a low approach over my head into Carlisle's little airfield, failed to get down across the short grass run, opened up and went round again. I watched the pilot's second attempt more carefully, noticing his hair blowing in the opened cockpit – no helmet. The same thing happened again; at about 10 ft the flaps were whipped up, the Battle sank almost to the ground and again roared into the air. On the fifth attempt it reached the ground and stopped, undamaged. It turned out to be an RAF aircraftman from York who, fed up with not being selected for pilot training, had decided to 'show them' on the spur of the moment. He had never controlled an aeroplane before and once in the air had not surprisingly lost sight of his own airfield. He had also forgotten to remove the pitot head cover and so had no airspeed indicator! He reckoned he must be just north of York so set his compass to south only to arrive at Carlisle, far to the north-west. Of course he was immediately arrested, as any encouragement of this sort of enterprise would have led to an airborne armada of frustrated pilots-to-be. Later I was glad to hear that he was to be given a pilot's course.

In September we had leave and went south to stay with Graham's parents near Redhill. Each day wild vapour trails twisted and turned in a soft blue sky filled with sound. I watched a Hurricane glitter in the sun, diving. It never pulled out, vanishing behind trees followed by the high scream of its runaway Merlin. Next day a Heinkel III crash-landed in a nearby field, its crew soon rounded up and removed. I sat in the cockpit, peering through the cracked perspex of its rounded nose. The rear gunner's position held empty shells and blood. Each warm and sunny day of that leave we stared into the sky, trying to identify Hurricane from Messerschmitt. Sometimes they were so high we could see only vapour trails slowly drifting above the rattle of machine gun fire and the whine of engines. Much later it was all called the Battle of Britain.

When we returned north I heard that the Air Transport Auxiliary (ATA) was taking women pilots to ferry light aircraft. If only I could get in and fly again; but the minimum experience for applying was 150 hours and I was not at all sure I had that many. I worked through my log-books, scratching together every minute in the air, trying to remember if I had forgotten to enter any flight, adding in the gliding minute by minute, and reckoned that some of my flying as a passenger had definitely been good experience. Altogether over the five years it came to 100 hours – plus a little imagination. I applied. In November I was summoned to Hatfield aerodrome for a flying test. I could scarcely breathe while I read the letter. The war was just beginning to get serious and I had to be involved; and it had to be in flying. Nothing else could even be contemplated.

4 · Early days in ATA

I reported to No. 5 ATA Ferry Pool at Hatfield on a cold 11 November 1940 to find the aerodrome and buildings dulled with camouflage and netting. After a brief, apprehensive interview with the famous Pauline Gower, Commandant of the unit, I was handed over to Margaret Cunnison for a flight test in a Tiger Moth. I had brought my own helmet and goggles, but was lent a sidcot suit to keep out some of the cold, and for the first time wore a parachute. By now I had not flown for over a year and was anxious that I would forget everything and botch the flight; but as soon as I started to taxi out it felt all right. The take-off was smoother than I expected and a little confidence returned, or maybe it was just my unconscious self pulling out all the stops knowing that my conscious self would be for ever unbearable if I failed. Fortunately, I did not. Margaret climbed out and said 'That's all right. You'd better get back to the office and fix up about when you start.' Overjoyed, I almost ran there – to the anticlimax of matter-of-fact statements about security checks, and that I should hear sometime in about two weeks. I did, I was accepted, and must report to Hatfield on 1 December.

On 2 December, having arranged lodgings and been issued with my own leather gloves with silk inners, sidcot suit and furry inner, flying boots and parachute, I had two flights with Margaret 'to get my hand in again'. I started ferrying on the first reasonable weather day, 5 December, my charge being Tiger R5033 from Heston to Aston Down, landing on muddy grass between the half-built runway, covered with concrete-mixers, and the contractor's Rolls-Royce: 1 hour 20 minutes. Then it was flying on every day that the weather allowed, learning to find camouflaged aerodromes, avoid barrage balloons, and appreciate very fully that there was no one to look after me in the air but myself. By the end of the month I had logged 18 hours, as much as I had previously been able to afford in two years; and instead of scraping up money to fly I was being paid, as a junior second officer, £8 a week. I was given a uniform with wings, tailored to fit, a forage cap and a greatcoat; the airforce-blue shirts and black ties you had to buy yourself. I was very happy even though my first few months were no easy ride. The winter of 1940/1 was bleak, snowy and cold, and most of it was spent in open aeroplanes – Magisters, Fox Moths, Tigers and their near relation, the Queen Bee. This was a cheap Tiger built to be shot down on gunnery practice and it was flown from the front seat because the back was supposed to be filled with heavy radio. Since it had not yet been fitted the Queen Bee fell easily on its face when landing. But this problem was less than keeping out the cold, which after a couple of hours had discovered every chink between glove and sleeve or collar and helmet. Slunk low in the cockpit I flew keeping as still as possible less movement allow any freezing air to find its target. It was luxury to be given the taxi Puss Moth to collect some pilot in the relative warmth of a cabin, and look down on the night's frost still outlining the north sides of houses. Within days I returned from one of these trips to find everyone totally preoccupied with the disappearance of Amy Johnson while ferrying an Oxford from Blackpool. It was the first fatality in our pool.

One of the practices in the very early days of ATA was to insist on the new pilot following a more experienced one on first long trips. I hated this because I was so busy trying to keep with the leader that there was no time to navigate, so I profited almost nothing from flying the route. My first follow was soon after I joined, with a Tiger from a grass field at Wroughton in Wiltshire to Church Fenton, near York. In the late and freezing afternoon my leader got lost, flying around vaguely looking for landmarks with me in chilly pursuit. I was beginning to get low on fuel and therefore worried, but kept following as instructed. Then I saw an airfield almost below with circuiting Tigers; it was either fuel or follow so I broke away and landed. Shortly after my leader came in, discovered we were at Doncaster, also refuelled, and we went off together to reach Church Fenton just before dark.

After handing over our Tigers we got transport to York station, my senior going straight to the office of the Railway Transport Office (RTO) whose job it was to point itinerant soldiers towards the best train for their destination. Within minutes we were given a schedule, authorisation to travel, and advised where on the train was the best chance of seats. The sergeant on duty also gave us a welcome mug of tea and a quick warm at his stove. Many times after that I was to be thankful for this helpful system.

The next time I was told to follow, on my first trip to Scotland, it did not work either. My Tiger was only from Blackpool to Prestwick, so I went there from Oxford in the front seat of my leader's Tiger. We set off from Blackpool's Squire's Gate aerodrome on a cloudless winter day but with patches of low fog over the sea. For some reason never explained my leader stayed at 600 ft flying through the top of the chunks of fog – presumably for fun – with me dutifully on her tail. Then, over the Solway Firth she dived into another patch; except that this was a massive fog bank. Quickly I closed up to keep her in sight; then she vanished totally and I was left at about 500 ft with no idea of my attitude; except that it was the same as my leader's when she disappeared. But it certainly was not straight and level and my speed rapidly increased. With no practical blind flying experience I went on to instruments as I had read in a book. 'Keep your eyes on the instruments,' the book had said, but suddenly the instinctive need to look out was overpowering. I looked. Straight ahead were grey waves in plan view and very close. I do not know what control movements I made to

Early days in ATA

wrench the poor Tiger on to an even keel, but that is what happened, and I found myself flying level between the base of the fog and the sea. There was not 50 ft between them and they were joined by swirling stalactites of mist. I could scarcely breath for the pounding of my heart, and flew straight on a compass course out to sea until I felt able to think about doing something better. When my hand reduced its shaking I turned on a course which would take me back to the coast at an angle so I would not fly slap into cliffs. The first thing I saw in the weeping greyness was a large black rock, its head in the fog. Then I saw cliffs, turned slightly to go between the two and fairly soon came out of the fog into the dazzling brightness of a serene day. Thoroughly shaken I went back to Carlisle. If I had not looked out at that instant I would for ever be resting on the cold and muddy bottom of the Solway Firth. It was now too late to set off again for Scotland, so I stayed at Carlisle overnight, although Graham was no longer there having been accepted as a fighter pilot, and was at an Operational Training Unit (OTU) in Lincolnshire.

The next morning I flew to Dumfries where my leader had landed. In sunshine from 1500 ft I saw the black rock I had passed in the murk. It had a lighthouse on top of it, the bottom of which had been in fog! After this, ATA discontinued following; you had to do it on your own. This made me much happier even though I immediately disgraced myself by getting lost. It was in a Magister from Cowley, Oxford, to Shawbury, in February, and this day of rare sunshine must have made me careless for I mistook a camouflaged Worcester for Stratford-on-Avon. They are both on rivers but at the time the whole country was flooded and the rivers had disappeared. Unaware I flew on until, instead of the gentle undulating land around Ternhill and Shawbury, there appeared great black hills with snow on top. Frantically I searched my map, but I learnt that day that when you are lost you become remarkably incompetent at finding yourself again; you wishfully think landmarks to be what you would like them to be. When I failed to locate myself, and with the ground below becoming increasingly hostile, I decided that I must return south to country I might recognise, so I turned and flew back the way I had come; except that I did not even do quite that. Fuel now was a problem with no airfields among the hills below, so I planned to let one tank run dry and rapidly switch to the other. I practised moving my hand quickly to the fuel cock but when the engine did stop the sudden silence still took me by surprise, and it seemed an age before the lovely noise began again. I flew on. The land became flatter and an enormous bend in an enormous river appeared, but so confused was I that I did not recognise it as the Severn near Gloucester. By this time I was not only really short of fuel but would soon also run out of daylight. Miserably I decided that I must land in a field while I still had a little of both aids to survival – ATA did not mind where you landed as long as it was an airfield. There were good fields ahead but each one, as I came to it, was obstructed by tripods of poles to frustrate Germans. Finally, it just had to be the next field, obstructions or not; I *must* land while I still had power and could see. But there were no obstructions.

Suddenly I realised I was looking at an airfield. It was so close it semed to spread to the horizon on both sides; I was almost on top of it. Feeling suffocated by relief I landed. It was 15 minutes to last light and I had just five minutes of fuel left. It was Aston Down, still with its concrete mixers and boggy grass, but to me perfection, though I still did not enjoy telephoning Hatfield, even after the duty officer told me sympathetically that no one knows about navigating until they have been thoroughly lost. I hoped that this would be true, but it was not the end of my troubles. Two days later I had a misfiring engine on a Queen Bee and returned to Hatfield, then three weeks later I had engine failure on another Queen Bee. This I had collected from Witney, near Oxford, a few minutes after it had landed from a ten-minute test flight following a rebuild. Some twenty miles on my way the engine spluttered. I checked everything I could; not much on a Queen Bee. Then it stopped, started again, and stopped for good. Switching everything off I hoped I could reach South Cerney aerodrome, but it was not to be, so I turned into a nearby field. It was long into wind, but ploughed with furrows across the wind. I had to land into the wind because of its strength, but as the ground was very soft and muddy I hoped that running across the furrows would not do too much harm. I dropped the Queen Bee, slowly, tail well down on to the squashy earth, and gently it went on its nose. I climbed out surprised at what a long way it now seemed to the ground. The prop had been almost stationary when I landed and was lying neatly in a furrow, undamaged. Within minutes a squad of airmen arrived from the aerodrome, cheerful at this diversion from routine. They got the tail down, pushed the Queen Bee across the field, half lifted it out through the gateway, and pulled it tail first along the road with me clumping along behind in my sidcot and muddy boots. In the hangar the sergeant said, 'No damage, Miss; you were lucky.' So once again I telephoned Hatfield; reporting muddy but no damage.

The next day I was summoned to the office. 'Why did you report no damage?' Pauline demanded. 'That's what I was told,' I said, 'and I inspected the prop myself.' 'Well, when *they* came to inspect it they found two dents in the bottom of the cowling. You had reported the aircraft undamaged and it wasn't true. Why did you land across the furrows? Everyone knows that this guarantees going on your nose.' I was dismissed from the presence with a flea in my ear. What I did not realise at the time was that Pauline was fighting a tough battle with the RAF and the Air Ministry to get the women's ferry pool recognised as competent to fly operational aircraft, and by having to report pilots who lost themselves, or aircraft undamaged when they were, did not help her case. What more natural than to let fly at the culprit.

Fortunately, though I obviously did not know it at the time, my tribulations had come to an end, and I was able to go on ferrying without attracting any more attention to myself. As the days lengthened with the spring and the elementary training schools went into high gear to keep pace with the increasing ferocity of the war, so it was Tigers and Maggies from south to north, and often from north to south – like taking in each other's washing. But there was method in the apparent madness; we mainly took aircraft to safe distribution centres in Scotland, while aircraft coming the other way would be replacements for schools. Soon the days were long enough to get a Tiger from Oxford to Prestwick in a day – 4 hours 15 minutes' flying plus time for refuelling twice – and come back on the night train. ATA at Prestwick held six sleepers permanently booked for ferry pilots returning south. In the winter it had taken sometimes three to four days to get north, waiting for fog to clear, or flying the long way round the coast because of low cloud, rain or snow, but between 1 and 6 May I took four Tigers to Prestwick, coming back on the train three nights running. Arriving at King's Cross station was not, of course, the end of the journey. After an early breakfast in the station hotel, train to Hatfield, and van to the aerodrome there was just time to catch the taxi Anson to do it all over again. It was marvellous to have a sleeper on the night train but necessary to try to go to sleep quickly in the long northern daylight as we were always woken going through the Midlands in the dark by bombs, gunfire, shouting, or shunting. Sometimes, after a shorter trip, it might be a day train from Yorkshire, arriving at King's Cross, starving, near midnight, but the hotel night porter never failed to produce sandwiches and stoke up the smouldering lounge fire while I ate them. Then it was bed and sleep if the night was not too noisy. There was usually some sort of raid going on but we never went to the shelters. Flying every day made it essential to sleep at night or at least rest horizontal in a bed; to do anything else would have meant exhaustion and broken aeroplanes. It was a balancing of the risks, biased by optimism. One night a cluster of incendiary bombs hit the wall between my bedroom windows, but none came in. Thousands of people, of course, spent every night trying to sleep on the Underground platforms, whole families packed together leaving just a couple of feet along the platform edge for travellers like ourselves. The air was stifling and if there was time and the night was not too noisy I preferred to walk – even with my heavy parachute bag.

Every day ATA put a great deal of thought into planning how best to get pilots to their aeroplanes and retrieve them from their destinations. The adjutant at each pool received the day's quota of aircraft by telephone early in the morning and quickly had to work out how to achieve the maximum ferrying for the minimum taxi flying. The taxi Anson, which normally carried up to ten pilots – although I squeezed in as fourteenth one day – would go to factories which had batches of aircraft ready; the Puss Moth flew a single pilot for the odd aircraft in another direction, and the van would take pilots to closer aerodromes. The evening collection was more complicated because not everyone would have reached their proper destination, particularly in bad weather. Sometimes it was a matter of getting home by any means you could find – by hitching a lift with an RAF pilot to a more convenient aerodrome, or in the taxi aeroplane of another ATA pool. From there they might even give you an aeroplane which took you nearer your own base, if you were lucky; so it paid to travel light. Nevertheless, you could be stuck out for perhaps a week so this process could not be carried too far. Detachable shirt collars helped as they could be turned inside out and, as a last resort, back again. Some pilots seemed to need more home comforts than others. Audrey was one of these, and when she acquired a minute white Peke puppy decided to take it flying with her, travelling in her parachute bag. Someone told her that if fed on gin it would stay little. Within months it was the size of a young polar bear, so sadly had to be left behind.

One weekend that summer was like the fun flying of peacetime. Early in the war civil aeroplanes had been requisitioned, though not always moved away from their home airfields. Now, in 1941, they were taking up valuable hangar space and it was decided to shift them to where they could either be of more use, or be stored out of the way. Several of these little aeroplanes were in Wales and I was sent with Diane Farnell and a Leopard Moth on a perfect summer weekend to round them up. First was an elderly DH60 Moth AV991, but in better times G-AACY. It flew like thistledown after the Tigers. I took this to Porthcawl with Diane following in the Leopard. We then went to Penrhos, for her to collect something, and I picked her up from Llandow. Then I had a Tiger from there to RAF Valley in Anglesey, landing at Towyn for fuel and wandering across the beautiful Welsh mountains in the afternoon sunshine, far from thoughts of war. Diane flew the Leopard back to Hatfield, picking up another pilot on the way, and I returned from Valley by train. After this I had another session as taxi pilot, but now as P1 and not stooge, the name given to junior taxi pilots who had to allow the more senior pilot they collected to fly home. It was a sensible arrangement though not popular with any new pilot longing to stack up the hours. Apart from carrying the usual ATA pilots, this session included an engineer to be flown to a problem aeroplane and then taking him home again, a new pilot from abroad to be taught some navigation – this really boosted my morale; and a senior WAAF officer from Andover to Sealand in Cheshire for an official inspection, and bringing her back next day. Everything went well including the weather, and she tipped me 2s 6d when I delivered her safely home!

By late spring 1941 Pauline was making real headway with getting us more usefully employed – in spite of the fumbles of at least one of her second officers. New pilots were joining our pool, which made me less junior, and there were rumours that some of us would soon be moved to Hamble on the South Coast to start No. 15 ferry pool there. In August I ferried an Audax – sister to the Hawker Demons I had looked up to as giants at Biggin Hill in ages past, was given some unofficial dual in the Anson, and dispatched to White Waltham for a conversion course for Class II aircraft. If I passed, this would clear me to ferry all single-engined fighters without further instruction including beautiful Supermarine Spitfires, which I longed to get my hands on more than any other aeroplane.

My instructor at White Waltham was Bill Smith. He showed me over a great fat Harvard, gave me five dual circuits and bumps, got out and told me to do some circuits on my own. I did 2 hours 50 minutes' worth trying to get all the flying I could. After tea he gave me a check ride and I was qualified for Class II. I never flew another Harvard.

Very properly I was not handed a valuable Spitfire next day, but served a short apprenticeship ferrying Miles

Issue 2.

MASTER

Engine: Fuel: 87 octane (All Marks).
  Mk. I: Kestrel 30. 2-position mixture control. Single-speed blower.
  Mk. II: Mercury XX or 30. 2-position mixture control. Single-speed blower.
  Mk. III: Twin Wasp Junior. Stromberg Y9J carb. (NOT injection). Start and fly in Auto Rich. Cruise in Auto Weak. Use manual range at after part of lever travel ONLY TO STOP. Single-speed blower.
Propeller: Rotol hydraulic, or counterweight, constant speed.
U/C Operation:
  Normal: Hydraulic. 3-position selector should return itself to NEUTRAL.
  Reserve: Hand pump.
  Emergency: Select DOWN or NEUTRAL. Depress EMERGENCY VALVE PLUNGER (under safety catch near left foot in Mks. I & II: red knob on hand pump body in Mk. III) and use hand pump. Before using Emergency, attempt to lower Flaps by hand pump with U/C lever NEUTRAL. (U/C requires about 150 strokes).
  Indicators: All aircraft: Green lights, locked DOWN. Early aircraft: No lights, IN TRANSIT. Red lights, locked UP. Later aircraft: Red lights IN TRANSIT. No lights, locked UP. Warning horn not always fitted. Also mechanical indicators in wings.
Flaps:
  Normal: Hydraulic. 3-position lever should return itself to NEUTRAL.
  Reserve: Hand pump (see also U/C Emergency). Emergency: None.
  Indicator: Electric at top of dash.
Gills: Crank handle at bottom of dash. Indicator adjacent. Mks. II & III: carb. heat control below right side of windscreen. Mk. III: oil cooler control adjacent.
Tanks:
  One (35 gals.) in each wing. Plunger control cock (Down is ON) at pilot's right. Mk. I and some Mk. II have fuel pressure gauge (2/3 lbs./sq. in.). Most Mk. II and all Mk. III have fuel pressure red warning light.
Starting:
  Mk. I: hand only (starter magneto). Mk. II: hand or 12-volt direct (booster coil). Mk III: hand or 12-volt inertia (press handle to mesh and press booster button). All Marks: raise fuel pressure with wobble pump while doping.
Ballast: Only on Mk. I prior to No. T8343.

FLYING PARTICULARS

Take-Off: (Mk. II Static r.p.m.: 2600). Trim:

	Boost:	R.P.M.:	Mixture:	Gills:	Elevator:	Rudder:	Flaps:
Mark I:	+5.	2750.	Normal.	Open	Neutral.	Full Right.	UP.
Mark II:	+4½	2650.	,,	'Climb' mark.	,,	Full Left.	,,
Mark III:	+4.	2625.	Auto-	⅓ Open.	,,	Full Right.	,,
	(throttle to gate).		Rich.	(Oil Cooler: Open.)			

Climb:

	Boost:	R.P.M.:	Gills:	A.S.I.
Mark I:	+½.	2400.	Open.	150 m.p.h.
Mark II:	+2¾.	2400.	Watch temps.	150 m.p.h.
Mark III:	+1.	2400.	,,	150 m.p.h.

Cruise:

	Boost:	R.P.M.:	Mixture:	Gills:	A.S.I.:	Consumption:
Mark I:	−½.	1900.	Weak.	Watch temps.	185 m.p.h.	25 gals./hr.
Mark II:	−½.	1900.	,,	,,	195 m.p.h.	30 gals./hr.
Mark III:	−2½.	1900.	Auto-Weak.	,,	180 m.p.h.	35 gals./hr.

Stall: Flaps and U/C UP: 75 m.p.h. Flaps and U/C DOWN: 58 m.p.h.
Glide: Flaps and U/C UP: 110 m.p.h. Flaps and U/C DOWN: 95 m.p.h.
Approach and Land:
  Flaps: Effect: Max. speed for flaps: Final Approach:
  DOWN. Nose Down. 130 m.p.h. 85 m.p.h.
  Notes: Rather nose heavy on the ground. Before take-off ensure rear hood side panels latched. Both hoods should be closed in flight.

ATA Ferry Pilot's Notes for the Miles Master. This was all the 'instruction' one got, but they did the job well

Masters with Wasp, Mercury and Kestrel engines. On one of the Wasp trips I nearly ended my days in the Bridgenorth valley, a much-used bad-weather route to airfields on the Cheshire plain. Visibility was very poor with low cloud steadily pushing me down into the gloomy hollows. Suddenly I felt an overpowering desire to get the hell out, and with the thought turned back steeply. It was a Spitfire diving southwards which suddenly filled the angle between my wings and fuselage. I could see the stitching in the pilot's helmet. Then it was gone. I diverted to Pershore, until the weather improved, landing without flaps because they would not come down. My log-book mentions the flap problem but I cannot recollect it; it does not mention the near collision but memory is sharp. Later I flew the Master on to Ternhill, a place I never liked as the CO would not allow any of us girls in the Mess, and we were brought a poor lunch by a waiter in a white jacket to eat in solitary state in the Ladies' room.

Progress towards my dream Spitfire continued with a Fairey Battle in dreadful weather with an airspeed indicator which did not work, and then it was thought I might be allowed out with a Hawker Hurricane. Still in a bit of a daze through flying a new type of faster aeroplane every few days I collected my parachute, maps, handling notes, and night bag, to be driven to the factory at Langley in the van. There I was deposited with my belongings outside the flight office and left on my own. Cloud was low with the far side of the aerodrome invisible, and the cables of barrage balloons climbing, Indian rope-trick fashion, into the murk.

As soon as I had signed for the usual petrol and loose equipment, I hurried to the Hurricane to find to my horror that it was a new Mark II bristling with cannon and smothered with overload tanks – they might at least have given me an old one to practise on. There was still no chance of flying so I got in without the parachute and went through the cockpit drill until the hard seat edge became too sharp. By then, being in need of tea and company, I returned to the flight office and its weatherbound test pilots until the van fetched me back in a fog at 6 p.m.

The following day was the same, but this time two of us

Early days in ATA

were dropped at Langley, the other pilot being an old hand on Hurricanes. He jumped into his aircraft and vanished into the mist, between the balloons, and down the railway line. They collected me in the evening with my first Hurricane still unflown, but no one complained. ATA policy was that the pilot was responsible, and must decide when to go – or turn back – and this decision was respected.

Next day I was given a chit for another Hurricane but this time from Brooklands. We went in a bus, eight of us, hot and nervous, reading our notes all the way there. We arrived near lunchtime, and as the machines were not ready, went to eat in the old clubhouse. It was sad to see the bright cream and orange building a dirty grey with untidy netting draped over the sharp corners. We had lunch in the old lounge, the floor bare and dusty, the talk and laughter gone. Instead of Pimms we had strongly chlorinated water to wash down cold, stodgy cod. It was a relief when the telephone rang to say the Hurricanes were now ready. We each climbed into our aircraft and soon, with a great roar of Merlins, were taxiing out to take off and climb away over the camouflaged race-track banking. My Hurricane, bound for Scotland, felt stiff, new, and most expensive with its four cannon, but this time with no overload tanks. The ground crew grinned when I told them it was my first, and as they pulled the chocks away my thousand horses galloped off at what seemed uncontrollable speed in the vague direction of the take-off point, with me mentally trying to catch up.

Another Hurricane was taxiing just ahead, its pilot looking out anxiously from side to side: it was his first, too. He turned across wind and his shoulder moved in little jerks as he ran through the cockpit drill and fiddled with the tail-trim. His head turned once more as he checked all round that the sky was clear; then the Hurricane swung into wind and in a cloud of dust departed. When just clear of the ground the undercarriage retracted without any dip of the nose, and it climbed smoothly away. Very nice for a first. I turned into wind and opened up. Almost immediately we were in the air with a marvellous surge of power. Now undercarriage up. The selector did not move. A second and more determined attempt brought no better result, except that the effort made my poor Hurricane behave like a drunk porpoise in full view of everyone.

Before going on to the Spitfire we were expected to do a number of drama-free Hurricane landings, and I wanted to get three out of this trip. The lateness of the start helped, because it was not going to be all that easy to get to Scotland before dark with the stuck selector keeping my speed down to around 150 mph. But neither did I want to land back and say I could not get the wheels up without trying some more, so I flew on over Oxford and Worcester in sunshine, at intervals renewing my struggles with the selector, until forty minutes later I went in to Shawbury to refuel. I asked if anyone knew how to free the selector, and also could they see why there was oil over the cowling? Slowly they got to work. They were short staffed. The petrol bowser had to come from the other side of the aerodrome. They took off large pieces of cowling, then filters; poking their heads up into the mass of engine they pulled out more bits: my lovely Hurricane was being rapidly disembowelled. I mumbled about daylight getting short and having a long way to go. They smiled, enjoying it all. 'We're doing our best, Miss, can't have young ladies flying around losing oil.'

I left them to it and went in despair to the canteen and used the time to ring through to the station met office for an Actual covering south-west Scotland. They said a depression was moving in rapidly and advised me not to go far, if at all. This was good as it meant that the others would probably still not be too far ahead. Returning to my Hurricane it was a relief to see the cowlings now on and the tanks full. They said nothing could be done about the undercarriage: until it had been worked more everything was so tight that I was having to push against the locks as well as operate the hydraulics. If, however, the locks could be held open by some means, it was possible that the selector could be moved. After some head scratching, the engineer officer tied string to the lock cables, and told me to pull this with one hand and heave on the selector with the other. They asked me to fly round the aerodrome so that they could see if it worked. Once well clear of the ground I unhooked the string from the primer, where it had been hung for con-

Early days in ATA

venience, with my left hand and got a firm hold of the selector with my right, holding the stick with my knees. At that instant we hit a large bump, down went a wing, and I grabbed for the stick. The string fell out of sight below the floor. I yelled with rage, slammed the hood shut and sulked away northwards, legs still dangling.

The weather was now worsening, with Liverpool's permanent fog making the river gleam yellow. Dirty smoke belched from chimneys and trains. The tide in the lower Ribble was out and miles of mud added to the desolation. It was a relief to glimpse a shadowy Blackpool tower away to my left and to fly over green fields and small tidy villages once more. Then up the edge of Morecambe Bay with the sand bright and clean in spite of the lowering cloud of the approaching front. Although the Lake District mountains were crowned with cloud and mist lurked in the valleys, the coast was still clear so westwards I went over Barrow and north around Black Combe mountain, its head invisible. There were two aerodromes within reach before the weather shut right down so I made for the larger and nearer, Silloth, joining the circuit with two Hudsons. Wheels down – already down, curse them, flaps down, into fine pitch and I landed – about four feet too high. Nose up and still airborne, I sat mouselike, without moving, with stories of Hurricanes being good at dropping wings flashing through my brain, but it lowered itself with only a 'wump' square on the runway. Feeling mad with myself I turned off at the next intersection and watched the Hudson come in. It overshot, it bounced, it pranced off the runway. Each time it touched there were wild squeaks from the tyres and jets of smoke. It ended up in long grass three feet from the main road. My little moment was forgotten in the excitement.

The Maintenance Unit man, in Home Guard uniform, told me that two others of my lot had come in because of the weather, and had gone to the local hotel, so I waited for transport in the guardroom talking with its two policemen. They had lived locally all their lives, they said, except for an uncomfortable spell away in the First World War, and they made tea and looked after me as though I was a traveller from the other side of the world.

Early days in ATA

Next morning three of us from Brooklands sat on the tables in the watch office swinging our legs, talking, and waiting for the rain to stop. During the afternoon a few more aeroplanes came in from the sunnier south. Winnie Crossley and Lois Butler landed from Hatfield in a Dominie to collect Hurricanes for squadrons in Kent, and gave me the latest gossip, then soon afterwards the daily batch of new Hurricanes arrived, and by four o'clock there were fourteen of us waiting for the weather further north to clear. We drank more tea. Then, quite suddenly, the first gleam of clearing skies appeared out to sea on the western horizon. Immediately there was frantic activity, with everyone wanting their machine started first. The clean blue sky climbed rapidly overhead in the freshening north-west wind, sweeping away the dirty cloud, as the first Hurricanes took off with a sudden roar, turning out to sea. I was furious as latecomers had parked their aircraft in front of mine, and I could not even start my Merlin; until some of the ground engineers, sensing a bit of competition with their pals, pushed my Hurricane backwards on to a small rough patch of ground, whisked a starter battery from under someone's unsuspecting nose, and I was away.

After take-off I made a routine, and again unsuccessful, attempt on the undercarriage; then high in the clean air over the Solway decided to make one last real effort. By letting the seat down as low as possible I thought it might be possible to exert more power on the lever with my foot. So after a good look round I lowered myself into the depths of the cockpit, then, pulling my boot with one hand and pressing on my knee with the other I got my heel fixed behind the lever, held the safety-catch in my fingers and heaved. Nothing happened except that this time my foot jammed as well as the selector. I dislodged the boot by pulling my foot half out of it, and decided to abandon my fight with this wretched undercarriage, and raise the seat so that I could once more see where I was going. Then the seat jammed. I do not know if any other pilot passed by the wallowing Hurricane with apparently no one in it but I didn't care; I just wanted to see out. Suddenly the seat came up and I was popped hot and sweaty up into the sunshine with a staggeringly beautiful view of the whole of south-west Scotland in the crystal air. Away to the east the departing front merged purple with the dark mountains. Against it there were two whole rainbows, and everywhere around there were Hurricanes, light against the storm, black against the sun, all going the same way.

At Prestwick next morning I was given another Hurricane, V9633, to go south, which was lucky as the others mostly got Tigers. A sky of washed blue had spread over the whole country, and visibility was 50 miles everywhere, even around Liverpool. This time the undercarriage went up like silk and my purring Merlin floated me south, without need for maps, over a September countryside looking as though England was just a big familiar garden. Again I landed at Shawbury for petrol, splashing down the runway through shining pools of rain still left from the day before; then up again to cross the vale of Evesham, its trees bright with fruit, down low over the house where my parents were living, and then fast over the tip of the Cotswolds, to race along the valleys of Thames and Isis to the Berkshire Downs creamcoloured with corn. Far ahead was London sprawling pink and grey to the horizon, with my destination a small grass oasis among those houses. I landed this kindly Hurricane on the smooth grass of Hanworth and went to get some tea.

5 · Spitfire

Then came the moment I longed for. I was given my first Spitfire. It was a new Mark V$_B$, AA746, its camouflage paint unscratched, waiting at Lyneham to go to a squadron at West Hampnett (now Goodwood). I had read the Pilots' Notes, been warned about the ease with which a twitchy first take-off could be achieved by over-use of the sensitive elevator; but like everyone else I still twitched into the air. But once there the Spitfire's delight in flying gave me sheer pleasure; some aeroplanes are harsh or reluctant, others without character, but the Spitfire was perfect and when I came into the circuit at my destination, she floated back on to the ground like a feather. As I taxied in two flights of Spitfires roared into the air to turn south in close formation for a low-level sweep over France – maybe tomorrow mine would be with them – and then a tall Dutch pilot gave me a lift from dispersal to the watch office on his motorbike. Life was completely wonderful.

Next day I had another Spitfire, P8780, which was even better as I made no twitches on take-off. Then I was back to Masters and Hurricanes, and a couple of weeks later 30 minutes' dual on an Oxford to convert to Class III twins, and after that my first Swordfish, V4444. The eight-step climb up to the cockpit was like Everest, but it was a delight to fly, rising into the air like a great but noisy soaring bird.

To fly all the different aircraft in ATA we had Ferry Pilots' Notes, small cards held in a stiff cover by a shoelace. That was all, but they were brilliant. No long-winded instructions on how to fly, just important facts such as approach speed, how to manage the fuel, and emergency systems for recalcitrant undercarriages. Sometimes, given a new type with no spare time before last light one read 'engine starting' while being driven to dispersal; 'take-off procedures and speeds' while the engine was warming up; and things like flaps and landing speeds while in the air. It worked; but probably because we were flying every day and with little else to think about but the next aeroplane.

At Hamble our job was to clear the many little factories in the southern counties, taking aircraft to maintenance units for the fitting of guns and radios. We also on-ferried Fleet Air Arm aircraft to Gosport, Lee-on-Solent or Worthy Down after they had been brought to Hamble by pilots from more northern pools. When we arrived home from our day's ferrying we would fly these aircraft before dark the short distance to Gosport or Lee, and be fetched back by car. It was not necessary for our adjutant, Alison King, to allocate any extra aircraft. There were always enough of us on the look-out for parked Swordfish or Walrus as we landed back in the Anson; anything to get more flying. On the four short days 20–23 November 1941 I ferried 4 Oxfords, 2 Spitfires, 2 Swordfish, 1 Walrus and 1 Albacore. It was heaven, and it went on like that all through the winter; 5 Spitfires and 9 Oxfords in the five days from 5 December, and over Christmas itself, I had 1 Spitfire, 3 Oxfords, a Swordfish, and an ancient Blackburn Shark with an extraordinarily complicated method of engine starting. One day that winter several of us went in the Anson with George Dutton – the only male among us – to a meadow at Cowdray Park to collect American things called Vought-Sikorsky Chesapeakes. After a cunning sort of landing to avoid antiinvasion obstructions all over the field, we found the Chesapeakes neatly lined up with sailors to get them going. It was very cold without any shelter so it was no unmixed blessing that most of the engines caught fire while being started; a

WALRUS I & II

Engine: Pegasus VI. **Fuel:** 87 Octane.
Propeller: Fixed pitch.
U/C Operation:
Hydraulic hand pump at pilot's right, and 2-position lever marked "UP" and "DOWN." Always operate pump until it feels solid. Electric indicator on dash shows words "UP" or "DOWN." Indicator switch built into throttle quadrant. Warning horn sounds EVERY TIME the throttle is closed, and is intended to draw attention to the position of the U/C whether alighting on land or water. Press button at rear of throttle quadrant to stop horn.

Tanks:
One in each upper wing, 75 gals. each. Two "ON-OFF" cocks on starboard side. Fly with both ON. Electric contents gauge on dash.

Starting:
Manual inertia starter with doper and starting magneto in starboard side of nacelle. Slow running cut-out forward of throttle quadrant. Two men to lay down inside tail while engine is run up.

Control Locks:
Thumbscrew through rudder operating lever, stowed adjacent. Hinged bar at top of dashboard attaches to control column, locking ailerons and elevator.

Brakes: Pneumatic, hand control on wheel, differential on rudder.
Anchor: In bow compartment with cable stowed on drum.
Tail Wheel:
Tail wheel or skid is restricted to 40° movement either way, therefore make wide turns when taxying. Some aircraft have water rudder integral with tail wheel or skid which may be interlocked with the air rudder by lever near throttle. ALWAYS disconnect water rudder for ferrying by pulling lever BACK.

FLYING PARTICULARS

Static R.P.M.: 1950—2050 at Full Throttle.

Take-Off:
Boost:	Mix.:	Carb. Air:	Trim:	Tail Wheel:
+2. (Full Throttle.)	Rich.	COLD.	2 divs. Nose Up. (All Hatches closed.)	Steering Disengaged.

Climb:
Boost:	Mix.:	Carb. Air:	A.S.I.:
0.	Normal.	COLD.	75 knots.

Cruise:
Boost:	R.P.M.:	Mix.:	Carb. Air:	A.S.I.:	Consumption:
−2.	Not to exceed 2200.	Normal.	FULL HOT unless air is dry and over 15°C.	83−88 knots (2 knots slower wheels DOWN).	40 gals./hr.

Stall: About 53 knots.
Glide: Wheels UP or DOWN 65 knots. Avoid using power off glide. Use full HOT air to carburettor.

Approach and Land:
R.P.M.:	Mix.:	Carb. Air:	Final Approach:
1200—1500.	Normal.	Full HOT.	65 knots.

Notes: Two men required to hold up-wind wing-tip float if taxying in cross-wind over 10—15 m.p.h. Rudder and elevator very sensitive during take-off. Power-on approach with carb. heat on recommended because of tendency for engine to stop with throttle closed.

The Walrus was built like a battleship, but we never had the luxury of either two men lying inside the tail or running on our windward floats!

few burnt so well that the engine was unlikely to ever go again, but mine was better tempered. I flew it to Wroughton but I do not know what they did with it because we never heard of Chesapeakes again. In poor weather we had time to do other things – such as running along the peri-track trying to teach one of our two Polish pilots, Barbara Wojtulanis, to ride a bicycle. She flew Spitfires with ease but never mastered the gentle art of pedalling on two wheels. Just before Christmas weather stopped flying and we had gone into Winchester to shop while the cold front with its snow went through. Later Barbara and I were walking across the blacked-out cathedral close in darkness relieved only by the gleam of snow-covered ground under a now starlit night. Unseen around us were the tall buildings and it was eerily quiet, even our walking made no sound. Suddenly, in a high-up invisible room someone started to play Chopin's Revolutionary Study, the music crashing out of the darkness with overwhelming challenge. Barbara stopped, startled at this old symbol of Polish hope. We listened without moving until just as suddenly the pianist stopped and the silence once more closed around us.

We flew very little from runway airfields, mostly from pre-war flying club grass aerodromes, requisitioned racehorse gallops, or even from farms with holes cut in the hedges to make a long enough run, such as Marwell (now a zoo). This was so as to have aircraft assembly widely dispersed. Barns or small hangars and nissen huts were built in clearings and covered with camouflage netting to make the woodland appear undisturbed, and each assembly unit had its own small canteen (serving bright yellow custard), office and stores. Components were delivered from factories after dark by lorry. In summer these remote places were delightful, full of wild flowers and hundreds of rabbits, but in winter they were cold and muddy, with rain or snow dribbling from the sodden camouflage netting overhead. It was from a gallops at Chattis Hill, near Stockbridge, that we collected some of our Spitfires. The field was a narrow strip sloping uphill to the north, so there was always the problem of taking off into wind uphill, or downwind downhill. Sometimes one of us would decide to go off one way and have an exciting moment avoiding another pilot whose ideas were different. At these little places it was necessary to clear aircraft without delay as there was nowhere for them to go except outside, visible to any German who managed to look. There were also other reasons for flying them out quickly. In February 1942 the siege of Malta made the island desperate for Spitfires. From Chattis Hill we had to fly them to the maintenance units at Brize Norton or Colerne for their guns and radios, and from there they were ferried up north to be put on an aircraft-carrier. On reaching the Mediterranean they were flown off the carrier as soon as Malta was in range.* These Spitfires were scheduled P1W, meaning top priority—wait; you stayed with the aircraft until it was possible to go.

On 1 February three of us were taken to Chattis Hill by car to sit for two days looking at our Spitfires in their blue and tan desert camouflage bright against the slushy grey snow, while difficult questions were being asked in the House of Commons about their non-movement. The answer, of course, was simple: we could not see the trees at the end of the field. On the third day it was still snowing, but there was a rumour of a slight improvement. I reckoned that the rumour was about right with emphasis on the slight, and that the only way to be sure of reaching Colerne was to carefully plan a route so that I could follow roads and railways all the way; and stick to it rigidly, even round a double loop in the Savernake Forest railway line. When everything was ready I watched for the promised improvement as the trees at the end of the field appeared and disappeared in veils of light snow. Then they reappeared for a little longer. Visibility was now up to 700 yds and cloudbase, if you called the amorphous weeping greyness a base, about 300 ft, perhaps 400 ft. Now was the moment to go. My Merlin started with its comforting rumble and I taxied out at once, doing my checks on the way, to take off uphill towards the trees which were still just visible. A minute later the second Spitfire roared into the air, but the trees disappeared again for the third, so it was taxied back to

* Described in Lettice Curtis' book, *The Forgotten Pilots* (G.T. Foulis, 1971).

the hangar. My plan worked. I flew as slowly as possible, flaps down. It was not possible to fly with the wheels down on a Spitfire for extra drag as the undercarriage leg obscured the oil cooler. The weather did not improve, but neither did it worsen, and I picked my way along the roads and railway tracks at 140 mph, flying the last mile along a lane uphill to the aerodrome. At Colerne they did not seem to have heard about any Parliamentary panic, but this was now their problem. About 20 minutes later I was astonished to see our collecting Puss Moth edge out of the murk, flown by a new, junior pilot called Bridget Hill. As stooge she asked if I wanted to fly back but she deserved something for coming to get me in such awful weather. She had even brought some newspapers, so I sat in the back with my feet up on the front seat and read the papers. After a while I thought I had better look out to see how we were getting on, and was horrified to see just underneath a hedge with a cow on one side and an Oxford on the other. 'Where the hell are we?' I asked. 'Don't worry,' said Bridget. 'My parents live here, I know it well.' Quickly I averted my eyes from any more undesirable revelations and read about the war; it seemed safer.

Not much later Bridget Hill was killed as a passenger in a taxi Fairchild flown by a pilot from another ferry pool. He and another passenger also died. This happened soon after I had been at White Waltham on a short – and I think experimental – technical course to obtain better understanding of Stromberg downdraught carburettors and stopping American engines. That week six of us billeted at Cookham Village would, most days, be taken to Maidenhead by the ATA bus after flying and have a drink together in the Bear before catching the public bus onwards.

In less than a year I was the only one of us still alive – Graham Lever was killed in the Fairchild accident; Malcolm Grant, friend from happy Redhill days, died forcelanding a Boston after engine failure; Don Jameson had an engine blow up on him in a Tomahawk; Stanley Herringshaw lost an engine on take-off out of Fairoaks in a Beaufighter; and a new, junior pilot called Tim Corsellis, spun in a Magister near Carlisle.

Issue 2.

SKUA

Engine: Perseus XII. **Fuel:** 87 Octane.
Propeller: D.H. counterweight type, 2-pitch. Pull handle for fine.
U/C Operation:
  Normal: Hydraulic. Lever should return to NEUTRAL. U/C takes 45 to 50 seconds to go Up or Down.
  Reserve: Hand-pump.
  Emergency: Select NEUTRAL. Push in red emergency knob at pilot's right, after removing safety clip. This introduces separate pipe-line. Use hand-pump till green lights show.
  Indicators: Red lights, locked UP. Green lights, locked DOWN. Stalks in wing surfaces show position of each leg. Warning horn is operated by A.S.I. pressure line. Horn will blow if U/C is retracted below 85-90 knots and upper red light on U/C indicator will glow. Horn will also blow during take-off if U/C indicator is not switched on. Test-button for horn is on A.S.I. switch, on floor outboard of right foot.

Flaps:
  Normal: Hydraulic. Lever should return to NEUTRAL. Flaps move slowly.
  Reserve: Hand-pump. Indicator: Electric, on dash.
Gills: Winch handle, right bottom of dash.
Tanks:
Three in fuselage. One forward, 39 gals.; two aft, port 60 gals., starboard 60 gals. Run with all cocks on. Forward tank will empty first, when forward tank reads 20 gals., turn off and hold in reserve for landing. This must be done as a full forward tank with no passenger makes aircraft nose heavy for landing. Either port or starboard will then begin to empty. Watch your gauges. Always turn a tank off before it empties to avoid an air lock. Master cock operates electric master switch and should be closed after stopping engine. Target tower: each AFT tank 17½ gals. max. Use only to cruise (15 mins. each).

Starting:
Cartridge. To start cold, turn engine over at least two complete turns by hand (sleeve valves). Turn priming cock to "CARB." and gently pump carburettor full. Set throttle ¾-inch open. Turn priming cock to "ENG." and prime engine 6 to 10 strokes. Turn priming cock OFF.
Control Locks: At lower edge of dash. Centre the rudder whilst attaching.

FLYING PARTICULARS
Static R.P.M.: 2450 (approx.).
Take-Off:
  Boost: R.P.M.: Mixture: Gills: Elevator: Rudder: Flaps:
  +2¼. 2650. Rich. 2 turns open. Neutral. 5° left. 20°.*
  * If run is long and boundary obstructions are not high, take-off without flaps.
Climb:
  Boost: Pitch: Initial: Flaps: A.S.I.: Gills:
  "Rated." COARSE. 100 knots. *THEN* UP. 110 knots. Watch Temps.
Cruise:
  Boost: Mixture: Gills: A.S.I.: Consumption:
  −1. Weak. Closed if possible. 150 knots (175 m.p.h.) 30 gals. per hr.
Stall: Flaps UP: 68 knots. Flaps DOWN: 62 knots.
Glide: Flaps UP: 100 knots. Flaps DOWN: 90 knots.
Approach and Land:
  Flaps: Effect: Max. speed for flaps: Final Approach:
  DOWN. Slight. 175 knots. 75 knots.
Notes: This aeroplane is unusually nose-heavy on the ground. Land with stick right back and keep it back. Use brakes with greatest care. For taxying on soft ground or in strong wind, at least one man must sit on the tail-plane. Two on tail for run-up.

Notes for the Blackburn Skua used by the Fleet Air Arm. All RAF aircraft had airspeed indicators in miles per hour and naval aircraft in knots

In spite of bad weather and the short days I ferried 45 aircraft that February, with a Skua and a Gladiator for variety. Then I had a Fulmar but like its elder brother, the Battle, I had flown, the airspeed indicator did not work either. We also put in a lot of passenger hours in the Anson, sometimes being delayed because aircraft were not ready when we arrived. On many days there was little time to eat, but we were young and hungry, often cold, and we needed proper food. One day I landed to find the Anson ready to depart but having had nothing to eat all day they said that they would wait for me to get something from the NAAFI if I was quick. An airman lent me an enormous, elderly black bicycle and I pedalled off for a cup of tea and a bun. Returning at high speed I just clipped the edge of one of the flowerbeds beloved by RAF stations and went clean over the handlebars, landing miraculously on my feet, in front of 200 airmen who cheered delightedly at this unexpected entertainment.

When we did have time to stop for lunch it did not take long to discover where we could eat best. Navy stations were generally better than RAF ones and they had plenty of brown sugar to go with the coffee, but RAF Lichfield was exceptional. The cook had been chef in a top London hotel and no one inquired how he obtained strawberries and cream for lunch. The casino at Blackpool was also a favourite – for its 'unobtainable' steaks – and we were not too badly off back at Hamble because the Bugle Inn always had cold roast turkey; so we ate there quite often. In other ways Hamble was a good place; we could look out over the sea on every take-off and we were far enough from Southampton not to be unduly bothered by the bombing; although one night someone got it a bit wrong and dropped everything around Bursledon where several of us were billeted. With the noise loud enough to cause us to leave our rooms we were standing in a doorway trying to decide whether to go back to bed or to the small damp garden shelter, when there was a shattering explosion like a bomber going in with a full load. At that instant a sphere of ball lightning, cricket ball size, came through the glass window, bounced on the bed, bounced on the floor, up against the wall, back on the floor and disappeared. Three of us saw it but we could find no marks from its passing. Next morning we found shrapnel all over the drive, though luckily none had gone through my car's canvas roof. One day about then I came into Hamble with a Spitfire to find the entire place deserted until the head of our favourite mechanic appeared out of the air raid shelter. 'There's a raid on,' he said, 'but nothing's happening, I'll come and give you a hand.'

One lovely trip I had on a beautiful spring day in late March was to take Spitfire BL416 from Hamble to 130 Squadron at Perranporth on the north Cornish coast. Visibility was unlimited and the only navigation needed was to fly in a straight line to hit the coast north of Perranporth and then fly along the magnificent cliffs to my destination. I reached the coast at Bude, and suddenly realised that I would go over that huge expanse of sand and sea so well loved from my childhood summer holidays. But when I arrived I could not find it. Nothing was recognisable as I flew around looking. The miles of pale sand, the blue sea with its big white waves, and the dark purple rocks of Daymer Bay, had all disappeared. Then suddenly it all fell into place: Stepper Point, Polzeath, Padstow, Rock, Brae Hill. It was all still there, but it was tiny, like a little gem that could be held in the palm of my hand. All the brilliance of colour was the same; it was just the scale that had changed through my growing up. I arrived at Perranporth, approaching over the sheer dark cliffs with my Merlin rumbling gently, to land down the short runway. There was no other flying and no one about. I parked the Spitfire and found a Sergeant in a little hut to sign my chit and fix up a van to take me to Truro station for the night train to London. It was not yet crowded and the only other passenger in my carriage was a tired Czechoslovak flight lieutenant, stretched out asleep on the other seat. I was just about to follow his example when the guard came along. 'I'll see you're not disturbed by everyone getting in at Exeter,' he kindly said, and locked the door until that hazard had passed.

During April 1942 I was given another spell of taxiing on the Fairchild Argus, which carried three passengers instead of the elderly Puss which took only one; and on our Anson

Spitfire

AX360, which I loved. The Anson I must have been the world's gentlest and easiest aeroplane to fly; and quite different to the Airacobras we suddenly had foisted upon us in June. Each one flew quite differently, the cockpit was cramped even for me, and the engine had a reputation for blowing up without warning. We had no Pilot's Notes yet for this aircraft and culled information from a test pilot whose buddy had just been killed on one. We collected the Airacobras from Marwell Hall, near Winchester, dealing with the paperwork in a requisitioned mansion and taking off from a nearby field. It was actually three fields with gaps cut in the hedges to make the run long enough. My first Airacobra was to Lichfield and I arrived there on a beautiful June afternoon – but too late for a strawberry and cream lunch. Curving into the circuit I put down the wheels. Two stubs popping out of the wings showed me my main wheels were locked down, but the nose wheel stub did not appear. I pulled up the wheels and tried again; no nose wheel stub. The emergency actions produced nothing either. So now I could either raise the main wheels and belly land, which I was not too keen on as the engine was behind the cockpit and one sat on the long shaft to the prop; or I could assume that it was just a dud indicator and that the wheel was actually down. This did not fill me with much enthusiasm either as the nose wheel leg was very long and if it was not there landing at the Airacobra's unusually high speed of 100 mph could result in some interesting somersaults. I flew around thinking, then wrote on my map, 'Is my nosewheel down?' Flying low past the watch office I let it go from the window. If the leg looked forward of vertical from the ground, I reckoned it would be reasonable to land on it. I flew round again and saw people standing outside the watch office, one of them holding my map, but they made no sign. After a further run and no action I had another idea – any growing shortage of fuel always tended to induce them easily. Flying low over a nearby field I banked the aeroplane until I could see my shadow clearly. The nose wheel was down and it was forward of vertical. So I came in, careered down the runway, and taxied back to park. 'We got your message,' one of the watchers said. 'If you had come round again we would have given you a green!' I flew eleven Airacobras and was glad when they were finished, and so I think was everyone else.

On 23 June I flew my hundredth Spitfire; actually it was a Seafire, distinguishable as the airspeed indicator was in knots instead of mph. Flying these lovely aeroplanes was now so instinctive that one day I was thinking that it would be perfect weather for flying, only to realise suddenly that I was 1500 ft over Winchester in a Mark VB. Maybe we took them too much for granted, although they rarely let us down. We never opened the canopy for take-off or landing as this would have meant wearing a helmet and from Tiger Moth days we had grown tired of having our hair messed up!

One day I took a Spitfire into Brize Norton where the SFTS (Service Flying Training School), training future bomber pilots on Oxfords, was well behind its schedule as a result of bad weather. The instructors knew where their priorities lay and as soon as they reached the end of the runway overcame any problem of waiting for aircraft on the approach by simply opening up and taking off. Fourteen times I got down to 50 ft on finals only to have to go round again – but in a Spitfire I didn't mind.

That summer I was dispatched again to White Waltham for conversion to Class IV twin-engined bombers. These courses were still embryonic although later in the war ATA developed a much more comprehensive training system, even teaching ab initio pilots. Nevertheless, the simple system seemed to work, my conversion starting with an hour on a Blenheim with Chris Rambaut. We climbed to 11,000 ft to do some stalls – higher than I had ever been before in anything; I usually flew at only a few thousand feet, except once when I climbed a Spitfire to 10,000 ft in order to see the airspeed needle go beyond 300 mph in the dive. After stalling we did a few circuits, and I was sent off for 4 hours of them solo. Returning a whole Blenheim to dispersal was regarded as a pass. Next day I was started on Wellingtons. Again four or five dual circuits, concentrating on single-engined flying, followed by solo. After nearly three hours of single-engine circuits out of an 800-yd field

BLENHEIM

Engines: Fuel: 87 Octane.
  Marks I, IV & V: Two Mercury VII, XV or 25. Mixture control: OVERRIDE—NORMAL, then Manual range, OR 2-position, NORMAL—WEAK. Single-speed blowers.
Propellers: Counterweight: Mks. I & IV, 2-pitch. Mk. V, constant speed.
U/C Operation:
  Normal: Hydraulic. Pump on port engine. 2-position plunger selector with thumb-catch release to right of seat. Some have 2-position lever right of dash. All have 3-position hydraulic power plunger at pilot's right: DOWN on ground, DOWN for operation, OFF (midway) when airborne.
  Reserve: Hand pump.
  Emergency: None or CO_2 or cartridge. Control aft of hydraulic power plunger.
  Indicators: Green lights, locked DOWN. No lights, IN TRANSIT. Red lights locked UP. Warning horn. Also mechanical indicators.
Flaps:
  Normal: Hydraulic. 3-position plunger selector to right of seat: Always return to NEUTRAL. 3-position hydraulic power plunger: DOWN on ground, DOWN for operation, OFF (midway) when airborne.
  Reserve: Hand pump.
  Emergency: None.
  Indicators: Mechanical.
Gills: Manual or electric. Control to right and rear of seat.
Tanks: INNER (140 gals.) and OUTER (88 gals.) in each wing. Only use Inners for ferrying. Outers may be empty or not fitted.
Starting: 12- or 24-volt direct, or hand cranking. Dopers and starter mag. switches in nacelles. Some have booster coil switches (under flap in roof).

FLYING PARTICULARS

Static R.P.M.: Mks. I & IV: 2300 minimum.
Take-Off: Trim:
  Hydraulics: Boost: Pitch or R.P.M.: Mixture: Gills: Elevator: Rudder:
  DOWN. +5°. Fine or 2650°. Override (if Closed. ⅜" Down. Neutral.
  *(100 oct.: +9 and 2750 r.p.m.) fitted), otherwise Normal.
                                    [Flaps: Safety Speed:
                                     UP. 120 m.p.h.
Climb (Max.): Pitch or
  Boost: R.P.M.: Mixture A.S.I.: Pitch: Boost: A.S.I.:
  Mks. I & IV +3¼. Fine. Normal. 120 m.p.h. *THEN* Coarse. 0. 140 m.p.h.
  Mk. V.: +3¼. 2400. Normal. 140 m.p.h.
Cruise:
  Hydraulics: Boost: Pitch or R.P.M.: Mixture: Gills: A.S.I.: Consumption:
  OFF 0. Coarse or 1900. Weak Watch 180 30 gals.
  (Midway). (if automatic) Temps. m.p.h. eng./hr.
Single Engine:
  Boost: Pitch: Mixture: Gills: Dead Prop: A.S.I.:
  Level: +1¼ ⎫ Coarse (above 1000 ft.) Normal. Closed. Coarse. 115-120
  Climb: +3¼ ⎭ Fine (below 1000 ft.). m.p.h.
            OR Mk. V: 2400.
Stall: Flaps and U/C UP: 70 m.p.h. Flaps and U/C DOWN: 60 m.p.h.
Approach and Land:
  Hydraulics: Flaps: Effect: Max. speed for flaps: Final Approach:
  DOWN. DOWN. Slight. 120 m.p.h. 90 m.p.h. (at 15,000 lb.).
                                                                  85 m.p.h. (at 13,000 lb.).
  Note: Weight varies considerably with equipment; check when collecting.

Notes for the Bristol Blenheim. The flaps-up safety speed was the same as the maximum speed at which flaps could be lowered!

in no wind I reckoned I would never be able to walk straight again. But I needed to go straight to the Wellington for the final check after tea that let me return to Hamble. I liked the Wellington. It was a friendly aeroplane although it flew as though a bit loose in its joints, presumably due to its stretchy geodetic construction.

Throughout that critical summer of 1942 the flow of aeroplanes increased, and in the light evenings we could well arrive home in the Anson to find a further programme made out for us. So off we would go again, hoping that there would still be some turkey left at the Bugle when we finally returned. One evening we went to High Post on Salisbury plain for some late Spitfires. Mine was still airborne on test when we arrived. It landed, the pilot got out and I got in without the engine being stopped to taxi out to the end of the grass field. As I turned into wind, the entire spinner assembly from the propeller fell to the ground, so I taxied back to the sheds one-handed, holding it outside the cockpit, to ask if someone could kindly fix it on a bit better.

During this same summer Graham had been flying low-level sweeps almost daily over France – also on Spitfires, though I never had the fun of ferrying one to his squadron. He had been shot down twice, once landing without all his tail and once having to bale out, but he never got hurt. Now with a DFC he was being posted to Staff College, first in Britain and afterwards in the United States. Already some ten years older than the average fighter pilot he was unlikely to go back on operations, so it seemed a good time to start our family.

1946 the WEIHE

6 · Peace freedoms – and problems

After I left ATA I went to stay with my parents near my Father's Royal Engineers base in Warwickshire. Then he was retired from the army due to his age and we moved to my in-laws' house near Redhill, while they moved to another relative's house in Gloucestershire. I missed flying but bringing up our baby daughter, Vivien, during the war left little spare time, particularly under the flying bomb route to London. The first V1 was a surprise as, apart from rumours of secret weapons, there had been no warning; suddenly there was this strange engine noise and an odd-looking aircraft flew over the house at about 100 ft. A mile further on the engine stopped and it dived straight into the ground and exploded loudly. That must have been the secret weapon! After that these flying bombs came over in a fairly steady stream, and I ceased counting the bangs when I had seen 50 actually explode. Some were not so far away bringing down a few ceilings, but with the heightened awareness easily produced by such goings on, I soon discovered that about 10 seconds before the engine stopped it very slightly changed its sound. If this did not happen until the V1 was past the overhead point it was not necessary to take shelter – with twenty to thirty of them rattling noisily over each day it would have been intolerable to keep diving into a cellar clutching a yelling baby. Then the army set up a row of balloons along our ridge with detonators on the cables; one swung over our heads when we picnicked in the garden. They never caught a V1 anywhere near us though they must have petrified chasing Spitfire pilots; one turned back faster than I thought was possible. Then the V2s started, with the explosions on Redhill aerodrome and in Nutfield village bringing down a few more ceilings and bursting open all closed doors; but that was the end. As the secret weapon sites were overrun, D-Day came, and our second daughter, Elizabeth, was born shortly before my Mother died tragically in her fifties – far too young for such a lovely person. Life became just routine ration cards and queues, until the end of the war brought thoughts of once again being able to look forward to a future.

Graham, though still in the Royal Air Force, wanted to get the Redhill Flying Club going again and I wanted to do the same with the Surrey Gliding Club, but there were many frustrations. The airfield was still requisitioned, the hangars full of now surplus stores. There were thousands of Tiger Moths going cheap, but their disposal was slow, and petrol, particularly for cars, was still rationed. Gliding clubs were even worse off because there were now almost no gliders at all; just the remains of a few troop-carrying Horsas suitable only as chicken houses. Nearly all the pre-war club gliders had been requisitioned and had now been broken or destroyed, as my little Grunau had been, or they had become rotten as a result of damp storage. All we had was our determination to start flying again – and a certain ingenuity well developed during the last five years to overcome shortages!

One of the first clubs to get moving was the Cambridge University Gliding Club, which had managed to secrete the famous Rhönadler, built by Hans Jacobs in Germany in 1935, and in which Eric Collins and John Fox had beaten British Distance Records before the war. They still had no airfield so over the chilly Easter of 1946 they took it to Rearsby in Leicestershire to do some flying. I went there, not by car, but in the Redhill Flying Club's new Auster G-AGVJ, which it could not use until it got its airfield back, and which was temporarily lodged at Kenley. VJ was most welcome as it had enough petrol to give some aerotows, though now with a rope of sisal instead of the fearsome steel 'snatchers' of pre-war days. In return I soon found myself airborne in the Rhönadler which creaked slightly as it flexed its old joints. After I landed a Cambridge student from Ceylon, Ray Wijewardine, took off and after release appeared to indulge in some aerobatics with the Rhönadler descending in great swoops. At the perigee of one of these the glider coincided precisely with the ground in a nearby sloping field and stopped. 'Idiot,' someone said as we trudged across the rough ground. 'What's he playing at?' Ray came to meet us looking paler than was natural and quite speechless; so we went to see for ourselves. The Rhönadler's canopy had disappeared, the instrument panel was hanging loose, and there were compression shakes along both leading edges between every rib. The ply and glue had failed and the disintegrating wings had remained attached only by their tough but now willowy spars.

A few months later another stored glider, a King Kite, shed a wing soaring over the Long Mynd and the pilot was killed. It seemed that even the few gliders we did have could be unsafe, as no one knew anything about the ageing properties of glue.

Shortly after returning home from Rearsby I had the chance to fly Doc Slater, Editor of *Sailplane and Glider* magazine, to Salzgitter in Germany so that he could write about the new RAF gliding club there. Naturally I immediately accepted, although civil flying was still forbidden on the Continent and to go I had to become an Accredited War Correspondent – even though there was now no war. Vernon Blunt, who owned the magazine, somehow made the necessary arrangements and I sewed the green flashes on to my old ATA uniform. Doc and I took off on a brilliantly clear May morning in VJ into a howling anticyclonic east wind, clearing Customs at Lympne and inwards at Ghent. There we had the luxury of two fried eggs – it had been only one for years – went on to München Gladbach and Gutersloh for fuel, and finally reached Salzgitter in the early evening. Even flying low it had taken us almost eight hours against the shrieking wind.

Salzgitter, in the charge of Squadron Leader Stan Haynes, was a dream. He, with RAF and local German enthusiasts, had rescued and repaired all the gliders they could find. The buildings were undamaged, the food adequate, and the Germans happy to have something constructive to do. They pulled VJ up the hill face with the glider retrieve winch and securely tied her down.

For the ten days we were there the weather stayed fine and hot, and as soon as Stan discovered that I had not yet completed the Silver C I had started with the five-hour flight in Dunstable's pre-war snow he set about overcoming this deficiency. But in spite of the sun the cloudless anticyclonic air was so stable that thermals were dying at a mere 2000 ft above the ground which did not leave much margin, particularly as they were rough and the sink between them strong. Finally, on 9 May, Stan got fed up with waiting for any improvement and sent me off in his bright red Minimoa for my 50 km distance. Circling in every turbulent little blue thermal before it broke its head against the inversion I just made it, landing in an enormous field at Lüthorst and finding myself quickly surrounded by barefoot German farm workers and the inevitable hundreds of children who pop out of the ground whenever a glider lands. Now there was just the 1000 m climb to do for my Silver, but the inversion was still below that height. To have the best chance Stan gave me the slow, lightweight Mü 13, and I flew this for two hot days from very low winch launches with the inversion still on the winning side. In the end, with our return to England imminent and Stan's determination to finish my Silver unabated, he decided to aerotow me in the late afternoon, when the inversion was likely to have been pushed up a little by the day's thermals. He would release me at 300 ft over the middle of the town, he announced. My streak of caution rose rapidly to the surface but Stan would not be put off. 'There's always a thermal there,' he said. And there was; the inversion had risen a little as predicted, and half an hour later I had hoisted myself up from the chimneys, and circling interminably in the tops of thermals just got the height with 30 m to spare.

Peace freedoms — and problems

The next day the faithful VJ was lowered down the hill face to the landing field. I paid my mess bill with 200 cigarettes, and Doc and I took off, refuelled from cans at the Commander-in-Chief's strip at Bad Eilsen, flying again via Gutersloh to Ghent. There I heard that signals were going out all over about some civil aeroplane flying over forbidden territory. Knowing how easily communications can fail and how difficult they can be to unravel, we wasted no time eating any more fried eggs and departed hastily across the Channel; but I never heard anything more.

Back at home there was still no fixed date on which Redhill would be returned to us so we flew from Kenley with VJ. One delightful week was spent on top of the Long Mynd in Shropshire, mostly with members of the Cambridge club; some of the time being occupied by Walter Morison shooting rabbits to supplement our still meagre rations. When this diet palled I flew VJ 150 miles back to Kenley, bought a load of unrationed haddock in the town and flew it back to the Mynd in time for supper: it was less difficult than getting car petrol to go to the nearby town of Shrewsbury.

At Kenley the revived Surrey Club could exist only in a small way, so we had just five members: myself, Lorne Welch and Walter Morison, both ex-RAF and lately back from Colditz, and Hugh Kendall and Anne Blackwell (later Kendall), both ex-ATA. But we were only a gliding club in name until we got a glider as well as an aeroplane, and this came about in a somewhat unexpected way. The first postwar meeting of the British Gliding Association had been held on 1 March 1946 under the chairmanship of Professor Brunt of Imperial College, and among various changes we decided to introduce a Research Committee. It was not only that we wanted to find out more about thermal structure and standing waves, but it seemed the best way at the time of obtaining government help with the acquisition of ex-German gliders, ex-balloon winches, and other badly needed equipment. It worked, and the first response from the Ministry was to give our Committee the job of allocating a small number of ex-German gliders to clubs, and for them to carry out handling and performance tests as a form of payment. This was how we acquired our Weihe. Now all we had to do was collect it.

The Weihe was at Cambridge so Lorne and I went there in VJ only to find it sitting on its skid, the droppable undercarriage having disappeared. Rumour had it that the previous pilot had failed to release the wheels at the proper time, just after take-off, and that they had subsequently fallen off high above the city centre. Not unnaturally lips stayed tightly sealed waiting for a storm to break, but nothing happened. Presumably they still repose in some ancient dusty attic and will one day come to light as a curious relic. But this did not help us take off a heavy glider with VJ's 100 horses, however hard they pulled. All that happened when we tried was that the tow rope stretched, the Auster danced nosily, kicking up dust at one end, while the Weihe remained totally immovable at the other. Soon a crowd collected and were persuaded to pull on the tow rope to help the poor Auster. Very slowly the Weihe began to move, its acceleration cancelled as helpers one by one prudently abandoned the rope. It took most of the aerodrome to get enough speed to loosen the earth's grip and fly. Returning home over the eastern edge of London to Kenley I was very happy; we could now fly every day either in the Weihe or towing with VJ – except for one very small problem: we had no trailer. So we would just have to make any outlanding in a field big enough to tow from. Of course, one day it was not; Lorne, unable to get back to Kenley, put the Weihe into a little sloping paddock. The farmer, with some doubt in his eye as he studied the 18-m span Weihe, acceded to our request to borrow his 2-m long flat trailer, normally used for carrying milk churns. We were not without doubts either until we remembered seeing a telegraph pole lying on a grass verge. Convincing ourselves on no evidence that the Post Office would not be putting it up that day we borrowed it, and bolted it between the tow bar and the little platform over the axle. Now we had a trailer long enough but with very quaint cornering characteristics, the wheels being virtually at the back end. But there was little traffic on the roads and with plenty of straw padding we got the Weihe back to Kenley in two journeys; the only pause in the proceedings being a somewhat hilarious few minutes when the wooden floor boards of my car caught fire and all was saved by a bottle of fizzy lemonade. It was long after dark when we dismantled our trailer and returned the original to the farmer and the telegraph pole to its grassy resting place.

But although we were having fun with our beautiful Weihe, a handful of German gliders and a few pre-war veterans of doubtful structural integrity were not enough to put the gliding clubs back in business, so it seemed like a miracle when a furniture manufacturer, Horace Buckingham of Elliott's of Newbury, decided to build 100 Olympias. During the war his factory had produced aircraft components and he wanted a way of keeping his skilled workforce together until controls were lifted on the purchase of timber for furniture. The Olympia, based on the 1937 Meise of that fine designer Hans Jacobs, had already been anglicised by Dalrymple of Chilton Aircraft. Then Dalrymple had been killed in a Fieseler Storch crash, and it was to the eternal credit of Buckingham that he took over production of this excellent glider; and to our delight that he gave the testing contract to us as BGA No. 1 Flight Test Group – which we had formed in order to fulfil the Weihe contract.

It was, needless to say, in the freezing winter of 1947 when we made the first flight on the Elliott prototype, Lorne flying the glider which I towed with VJ. Apart from the weather we had few problems, the general handling, spinning, and terminal velocity dives with airbrakes open all being more than acceptable. Being the lightest pilot in the group I had the c.g. aft spins to do, and the required five turns ended with an entirely normal recovery. The Olympia was just what we all needed. It was pleasant and safe to fly, and it soared well.

As spring came and slid quietly into summer we tested the bright red, blue, cream or green production aircraft which Buckingham was now turning out with no greater difficulty than tables or chairs. Most of this flying was done from the disused RAF airfield at Welford. Buckingham's men would bring the gliders there in trailers, and later return for more, taking back those we had flown and passed, while Lorne and I would fly over from Redhill – which had at last been handed back to Graham – taking it in turns to tow or be towed to 3000 ft. From this height we could fit in the test schedule, including opening the air brakes in a 130 mph dive, if all went well; if not we made a further flight. One day in June we tested eight new Olympias plus one retest; I did five and Lorne four. Elliott's men were as happy as we were. In the long light evenings we would fly back to Redhill in VJ at peace with the world, looking at the lengthening shadows over the Downs and the changing colour of the fields, but ravenously hungry.

With the Surrey Club now at Redhill we needed to get down to instructing once again. Everywhere in Britain this was still done solo; using Slingsby Cadets which had been built for the Air Training Corps, because there was no suitable two-seater; but it was unsatisfactory, and there was a great need within the BGA to produce both an instructional policy and training standards for the future. At Redhill we were fortunate as Lorne, with 1400 hours' RAF teaching on Tiger Moths, had taken on the job of Chief Instructor, and we were soon training every day, now joined by the Imperial College Club; but best of all, with VJ as a towplane, we could concentrate on cross-country soaring. By midsummer we had more pilots leaving the airfield on Silver C attempts than we had earlier dared to hope. We had half the airfield, sharing with the RAFVR and the flying club, which Graham was getting back into shape; but even more important, we had plenty of members. Many, like Hugo Trotter and Jack Karran, were wartime pilots who now just wanted to enjoy their flying, while others were beginners wanting to learn. With such a wealth of talent and enthusiasm it was not surprising that the Club soon developed a momentum which carried it forward at exciting speed. Looking back it is difficult to believe how much

Retrieving the winch cable by Beaverette

Peace freedoms — and problems

happened in so short a time, in spite of all the shortages. For retrieving the winch cable we used cut-down armoured cars called Beaverettes, which Jack Rice at Leicester had winkled out of some Ministry. Lorne designed a winch which members helped to build. We had our beautiful Weihe, two Cadets, and had ordered three Olympias. And in the winter, with the thermals in hibernation a party of us from the Club went skiing, not now to Germany but to the high mountains above Zermatt. Each February the Surrey Club ranged over the Theodul glacier on expeditions (no lifts or snocats) or raced down from the Gornergrat to gain our Ski Club Silver for speed; I even rose to the height – or depth – of gaining soft snow Gold. Neither did we miss any embryonic après-ski party that was going, appearing at one fancy dress ball as deep sea mountaineers; eight glider pilots in bare feet wearing swimsuits or other brief costume and roped together, causing consternation on the dance floor.

There was something delightfully carefree about those early post-war years. Skiing was still cheap, and in flying there was no air traffic control, no permissions to land needed almost anywhere, and very little paperwork. The unwinding of war was taking a long time and as long as one kept clear of any intermittent bursts of officialdom life was pleasantly uncomplicated.

In June 1947 the BGA held the first postwar National Competitions at the Royal Naval Air Station, Bramcote, and of course we took VJ, the Weihe, now with a bright red fuselage, and an Olympia. It was not a very serious contest, more of a great meeting of old friends, and a cautious amicability from the Navy who were not too sure about our sort of aviation. VJ did some of the towing, to supplement two Tiger Moths, and we won a small collection of daily prizes with Philip Wills winning the competition overall.

One visitor to Bramcote was a Swiss pilot, Hans Würth, with a slow and elderly two-seat Spahlinger S-25, who, on the last afternoon, asked me if I would like to fly with him. As the weather had been poor all day, with landings less than 10 miles away, I accepted, thinking that we might do around 15 minutes' local soaring before packing up to go home. But as we released off tow after tea the cloud dispersed and the air became very slightly buoyant. With the utmost care Hans worked every weak thermal he could find until it faded to nothing at 3000 ft while we drifted on the gentle summer breeze over Lincolnshire fields of rich green. The thermals were widely scattered but Hans found them instinctively, almost as though he could see them; it was magic, and I sat in the front seat marvelling at this blend of superb technical expertise and the beauty of the land over which we passed so silently.

It was almost eight o'clock on this June evening when the lift finally left us in its endless pursuit of the sun, and we returned to earth. Hans landed on a disused airfield just as gently as he had flown. It turned out to be Full Sutton in Yorkshire, 125 miles from Bramcote; but there was no need for me to interpret for this Swiss as we had arrived in a POW camp still containing Germans. Willing helpers moved the glider to a hangar, we were given supper, and, after telephoning an astonished Bramcote, were put up for the night.

Even more unexpected than this delightful flight was my next meeting with the Spahlinger. After Bramcote, Hans had left it at the Derby and Lancs Club for members to fly and now it had to go home to Switzerland; could we tow it there with VJ? It was November with short days and temperamental weather, maps were still difficult to get, and few aerodromes in France had fuel. This was our biggest headache as with the glider on tow our cruise speed would be a slow 65 knots, which meant a pretty flexible refuelling plan. But these were the only problems. There was no controlled airspace or even many aeroplanes about. So as soon as a reasonable weather slot appeared we went; with the first stop inevitably Lympne for Customs. From there we set off across the Channel into a clear cold sky, arriving at Lille after 1 hour 50 minutes' flying. The aerodrome was deserted, both air and ground, so after seeing just how far I was going to have to taxi from the end of the runway I landed the Auster in the empty car park. Lorne, being more proper about these things, landed the glider on the runway, and in due course some people appeared, willing to push the Spahlinger in gathering darkness to a hangar.

Peace freedoms — and problems 79

Next morning we set off for Rheims as soon as it was light, but finding fuel there among the quantities of trucks and wartime detritus looked hopeless until we found a source of petrol in chipped and greasy cans which a French workman assured us was suitable 'pour tous des avions', and promptly poured it into VJ's tank using his beret as a filter. While I was refuelling, Lorne, who had landed the glider in position for take-off, collected the tow rope from where I had dropped it and laid it out ready to go, hitched on at his end. If anyone turned up to run holding the glider wingtip we preferred it, but if not we managed without. But Rheims had not seen the last of us as 15 minutes later we flew over widespread ground fog and returned. About an hour later we took off again, refuelled more easily at Nancy, and set off through the Vosges mountains – Lorne still maintains I went up the wrong valley – to reach Basle-Mulhouse in the late afternoon. Here the same bleak emptiness prevailed as I quickly taxied over to the dilapidated control tower to report; there was only about 40 minutes of daylight left and we still had to get over the Jura mountains before dropping down to Grenchen on the Swiss side.

'You cannot go tonight,' the man said, 'a Dakota is coming in soon.'

'But we are expected in Grenchen today, the glider is urgently needed.'

The man looked at his watch. 'How long to take-off?'

'Three minutes.' I smiled and he looked at his watch again.

'All right, but if you cannot go in three minutes you must come back.'

I shot out of the office, wound up VJ and fled across the grass. Lorne had laid out the tow rope and was standing in the cockpit talking to a man holding the wing tip. Swinging round over the end of the rope I jumped out, hooked it on, waved a sort of urgent signal, jumped in and with a quick glance behind opened the throttle. Lorne ducked down pulling the canopy with him; exactly three minutes.

In the grey and fading light I flew VJ straight for the 3000-ft tree-covered ridges, almost bending the throttle lever as I willed its struggling horses to get us over without the delay of having to turn away for a second bite. After an age the tops of the pines slid beneath, and I could start to descend as fast as I reckoned the feeble airbrakes of the old Spahlinger could manage. As soon as the airfield shaped itself out of the gloom Lorne released and we both dived for the luxurious prospect of a warm clubhouse and unrationed food. Snow came with the darkness and the blizzard went on for three days.

As soon as the weather allowed we started home, clearing Customs at Zurich and making Rheims that day, as without the glider we could cruise at 90 knots instead of 65. Rheims was just as cluttered but we tucked VJ into a safe place and got a lift into town to a small hotel recommended by the van driver. From there we went to look for a restaurant, but discovered that Bizet's opera, *The Pearl Fishers*, was on that night. Neither of us had seen any opera for years and, as it turned out, neither had the French as the house was packed except for a few seats right up at the back. The opera started on time but that is all that could be said about punctuality because the encores took over; the audience simply could not have enough. At 10.30 we were still at Act 2, with the singers in simulated leopard-skin trunks obviously happy to go on all night. They probably did but we left, worried about being locked out of our hotel. In the cold fresh air we found that neither of us could remember its name or how to get back to it – only that it looked out on a park-like green. It took over an hour to find it, but Madame had still not locked her door.

Next morning, again full of beret-filtered fuel, we made Lympne in just under two hours, to land back at Redhill just as the weather started to show more of its winter teeth.

But just as unexpected as the Spahlinger tow to Switzerland, No. 1 Flight Test Group were suddenly given a contract to test the Swiss Moswey III, so in early February Hugh Kendall and I set off back there in VJ. This time the weather deteriorated too fast for us and we landed in a French field just before the cloud hit the ground. It was good that it did not take long for the police to arrive as the field was too small for take-off with two up.

'Is there perhaps a larger field nearby, or better still, an airfield?' we asked them.

'Yes, there is indeed an airfield. It has not been used for some time but is probably all right.'

It was about 5 km along the canal and the police would be pleased to take me in their car. As the drizzly front passed Hugh started up and we watched him take off, clearing the hedge with little to spare. The police, determined to be in at the landing, roared off, foot hard down, along a road which every few hundred yards crossed and recrossed the canal via narrow right-angled bridges; priorité à droite had nothing on this. I hung on in the back as we screeched round each corner, but they saw the landing; Hugh having sensibly studied the field with some care before coming in.

In spite of this hiccup we made it to Birrfeld the same day to find the clubhouse technically closed for the winter and without heating. For the next few days we flew the little yellow Moswey, I doing the position error, trim speed and c.g. aft spins (again), and towing Hugh for everything else. On one take-off his airbrakes opened – and they were more powerful than those of the Spahlinger. In the tug the climbrate subsided to equal the rise of the surrounding countryside, in Switzerland a journey with no future. I had only two choices: to ditch the Moswey with all the complications of retrieving it, or to follow a nearby rock-strewn river-bed to gain a little time. It was probably the closeness of the rushing water that caused Hugh to wonder what the hell I was doing and to look around for salvation. Suddenly the airbrakes disappeared and VJ leapt up happily.

Test flying that winter did not even finish with the Moswey, as on our return to Redhill Elliott's had the first of their anglicised SG–38 open primary gliders ready for us. Lorne did the first freezing flight to 4000 ft which I towed, sitting in the warm tug and peering at the glider on the end of its tow rope through the roof, instead of the rear window; as the maximum permitted tow speed was 45 knots. Then we had Slingsby's new Gull IV for trials and some photographic trips with that wonderful photographer Charles Brown. On one of these sorties I was flying him in VJ when without warning he opened the door, leaning right out with his enormous and elderly camera to avoid having his foreground ruined by the Auster's struts. I hung on to him tightly in case he got any more ambitious ideas, as he was far too valuable to lose.

7 · High hopes in the Alps

Gliding has always possessed a strong competitive element, and in 1948 the great Wasserkuppe International contests of the 1930s were still a fresh memory despite five years of war. Now the world was again ready to compete in friendship, and where better than Switzerland, high in the clean alpine skies above Samedan. This first great meeting was not to be missed; and almost every pilot in Britain with some sort of experience, even some without, wanted to go. The problem was who to choose, as each country was limited to six pilots. So far we had had only our easy-going Nationals at Bramcote to use as a guide to team selection plus, of course, pre-war demonstrations of skill. This, naturally, put Philip Wills as number one, and it was easy to add Kit Nicholson and Donald Greig, both of whom had also flown in the war. Charles Wingfield, owner of the famous white Kite, Gracias, made up the fourth from British shores, while RAF pilots Jock Forbes and Pete Mallett would join the contingent from BAFO (British Air Forces Overseas) in Germany, flying Weihes. Elliott's lent two Olympias, Slingsby two Gull IVs, and the Rover Company five of their new Land Rovers. These were still being developed so the Swiss expedition would be considered as field trials; little did Rover realise how true this would be.

The BGA had preparatory meetings in London and when the matter of a team manager came up Donald Greig proposed me. I accepted immediately though little knowing how this small decision would affect almost all my future life. By the beginning of July we had everything ready. The Land Rovers were packed to the roof with spares, tools, and personal junk as we set off for the three-day journey. Over the cold 7500 ft high Julier Pass we drove into cloud, which occasionally broke to reveal snow-covered peaks seemingly overhead, and we began to realise that we had much to learn if we were to succeed in a country where it froze in July. But we were a happy team and our hopes were high.

Samedan, like most alpine airfields, was a single strip lying along the centre of the valley north-east of St Moritz with the wind blowing either down it or up it. When the sun shone it was enchanting, but with cloud covering the mountains the gloomy air had a claustrophobic quality.

The 1948 competitions were to be run by a young fair-haired Swiss pilot and lawyer called Pirat Gehriger, who had ambitious ideas about tasks for Speed over a Closed Circuit Course instead of the traditional Free Distance; and also about the use of radio for turning point observers. Launching would be by aerotow, using a Fieseler Storch, and by an electric winch; and as we rigged our gliders next morning in brilliantly clear weather at the start of the practice week everything looked good for us. It was Kit Nicholson who discovered the shortcomings of the electric winch, being launched at the same time as the local hausfraus were cooking lunch; the few volts left over gave him so little height that, half stalled, he released so as to land before he ran out of aerodrome. Off we went with a Land Rover to collect him, over the rough grass as the only runway was for some inexplicable reason under extensive repair, and promptly sank into two bogs on what was supposed to be the landing area.

A few days later the BAFO contingent arrived, bringing an additional Weihe, plus two jeeps and trailers full of stores. They also had radio in their gliders; as yet unobtainable in Britain; so we were now complete as a team. The pilots flew as often as the electric winch allowed, even if it meant landing out; though this was good practice, too, as the fields were tiny. Retrieving was even better practice – although more often considered a masochistic form of slavery – because with the tourist boom still far into the future, many of the roads were bottom-gear hills with such sharp bends that trailers had to be unhitched and man-handled around each one. It had also now become so cold, with frequent snowfalls, that everyone was wearing all their summer clothes at the same time.

Another unforeseen problem was that many fields, apart from being small and rough, contained invisible stakes for drying hay, which brought a real risk of glider damage. This quickly encouraged us to make the spare BAFO Weihe fully serviceable, though the work took longer than expected since it had been brought from Germany on an open trailer and was fairly full of rainwater, including the instruments, but good fortune had given us 'Pops' Kent, an ex-Navy engineer, as crew and he was only happy when working. With five days left to go before the official start on 19 July the British team got together to plan strategy, and the first decision we made was to put our gliders back in their trailers every night. There was only one hangar and Swiss enthusiasm for packing in more gliders than was possible, usually at the run, was too great a strain on our nerves.

Luckily, the next day the electric winch went up in a shower of sparks – we wondered if it could have been the iron that Joy Pressland was using at the time to help keep the team looking smart. It was replaced by two robust Ford winches, but until the contest began they could not show their worth as it rained every day.

The weather perked up for the opening parade of the eight countries represented, and the first task given by Pirat was an Out-and-Return race over the mountains to Davos, 34.5 km distant, with the Weisfluhjoch hotel as the turn point. Additional marks were to be given for the highest altitudes gained. All our pilots returned including Lorne – who had replaced Charles Wingfield who was ill – but it was the altitude part of the task that had the same fascination as watching a snake. Only one cumulus of any consequence developed, growing fast over the finish line on Muottos

Muraigl, so as each pilot returned in he went. On the airfield we watched, waiting for bits to fall fluttering to earth. Remarkably no one collided, though it led to Altitude quickly being eliminated as a points-scoring task! As might be expected other rules were made or rewritten as a result of this first championship, and one of these was to require pilots to be retrieved with their gliders. This time the trigger was one pilot who came back in the comfort of a private aeroplane while all the others were helping their crews manhandle the trailers high on the freezing passes in the dark.

The third task given was – unknown at the time – a historic one: the first 100-km triangle ever to be set and completed. All our pilots got round but the fastest was the Swiss, Siegbert Maurer, to become the first world record holder in his Moswey at a speed of 70 km/h.

One thing we could have done without at Samedan was trouble with the new Land Rovers, but back in England the company had discovered the need for a gearbox modification, and had sent out Peter Wilkes with the bits. Although only a small modification it necessitated removing the complete engine and we had no workshop. But we did have Ron Claudi among the crews and he built some shear legs in the open and worked non-stop with Peter. By the time they started on the fifth car they had got the job down to $2\frac{1}{2}$ hours. Each crew in turn had my Land Rover while their own was being done, and the only real problem we had was finally getting all the forgotten jackets, socks, and chocolate bars back to their rightful owners.

We were honoured at these championships by the presence of Lord Brabazon, Chairman of the Royal Aero Club and constant visitor to St Moritz for his beloved Cresta run, who one evening invited me to dinner at the Palace Hotel. I had not seen food like it for years, and I found myself observing an impressive creation of ice cream shaped like a church steeple, about 18 in. high, and smothered in chocolate sauce and thick cream. Then the waiter asked Lord Brabazon for our order. 'Lemon water ices,' he said, and I was too shattered to try to change his mind.

For the next few days the contests continued unevent-

High hopes in the Alps

fully. There were minor excitements such as the Spanish team captain inadvertently reversing his car over Egypt's wing; the Italian pilot finally ran out of cash and went home too proud to accept money from friends; a twin-engined Dutch aeroplane tried to land over the start line on top of the winch cables and on being waved away landed downwind on the cables straight towards the parked gliders; and a small Czechoslovak aeroplane came in with a family on board to ask for political asylum.

Then came the big task that everyone had been waiting for: each pilot would count the best flight he made to a goal of his own choosing on any of the next three days. On the first Ara Torrell of Spain set the pace by being the only pilot to reach his destination at Sion, 200 km away; our best was Philip at 92 km. Next day the weather was poor but it improved for the last day, 28 July, when all our pilots declared goals in Italy, except Kit and Donald Greig who chose Nice, via the Maloja Pass. While waiting at Samedan for news of landings we got on with various chores, such as the team accounts and washing a car or two and by mid-afternoon knew that Jock had reached Locarno, 97 km, while Pete, Philip and Lorne were between 79 and 41 km. It was nothing spectacular but Kit and Donald could still come up with something better. Later news of longer flights came: Pelle Persson of Sweden at his goal of Geneva, 295 km (which gained him the championships), and the Swiss pilot Schachenmann at Lausanne, 252 km. Still nothing from Kit and Donald. At 6.30 pm a confused telephone call told us that Kit had crashed near Chiavenna in Italy. I left immediately with his wife, EQ; Eddie Lauber, the Swiss team manager, taking us in his car. Half-an-hour later the emergency procedure for missing aircraft was taken for Donald Greig.

We arrived at Chiavenna Catholic Hospital in sweltering heat and darkness. Kit was not there. Eddie, who spoke Italian, tried to get information but nothing he was told seemed to relate to anything else. One person said that Kit was injured and would be soon brought in, and another that he was dead and still in the mountains. Then Philip and Kitty arrived to look after EQ, and Eddie drove them all back to Samedan in his car. I spent what was left of the night on a spare hospital bed, and set off early in the morning in Philip's Land Rover, with a team helper Alan Clarke, to see if I could find what had happened to Kit. We stopped at local police stations but soon found that the many small children wandering about seemed to know most. Eventually, and after a great deal of frustrating misinformation, we parked the Land Rover at the foot of a towering mountain and started to climb up on foot in the hot sun. About 1000 ft up the track we met a party of Italians coming down. They had Kit's body on a makeshift stretcher and told us where they were taking him. They also pointed high up the mountain with expressive gestures to indicate where the glider had crashed. With sadness we resumed our upward trek, as I had to do my best to find out what had happened as far as it was possible to do so. In about half an hour we came to a tiny village where people ran out of the rough little stone houses, chattering; but when we moved on they restrained us, pointing to one of the houses. Then, as a young man came out with an old rucksack on his back, they smiled; he was obviously to be our guide. This was as well because the track soon disintegrated into indeterminate sheep trails and then to nothing as we plodded even higher. Now it began to grow dark, although early afternoon, as a vast thunderstorm brewed up around us. It was only our smiling guide that kept me, at any rate, from being very frightened when lightning flashed close with sizzling noises. Then at 6000 ft in deep gloom we came to a huge mass of enormous stones, almost a mountain on a mountain, and as we laboriously clambered over the rocks I could see that Kit's glider had crashed on the very top. The shattered remains still formed the normal outline of the Gull. It must have hit square, fully stalled. Alan and our smiling guide wanted to go back now that we had seen the Gull, but I knew I must look it over carefully to try to find the cause. This was probably the only chance anyone would have; already the instruments had been taken or stolen, and obliterating snow could come at any time. It was not easy to inspect anything in the mix of near darkness and blinding flashes, but all the control cables were properly connected to bits of the appropriate surfaces,

1 AW aged 15 with Charlie Allen and his Puss Moth, 1932

2 AW's Royal Enfield motorbike with its infuriating circular fuel tank, 1933

3 Box Brownie Bulldogs at Biggin Hill, 1933

4 First solo completed: G-AAIM, AW and Instructor Nash, 1934

5 Brooklands between the wars

6 Anglo-German camp at Dunstable, 1937: first hops in unsprung Dagling

7 Dunstable 1937: AW splicing the pulley rope of the hilltop retrieve winch

8 Skiing above Berchtesgaden, 1938: falling about in the snow

9 Rossfeld ski hut. *l–r back row:* James Hayes, Alec Barnard, AW, Widdle Shaw, Jochen Benemann, 1938

10 AW and Grunau at Dunstable with family dog, Bogie, 1938

11 Surrey Gliding Club, Buckland, 1938: AW repairing broken ribs on the Grunau

12 John Sproule, in waders, instructing Arthur Saville on the Dagling at SGC

13 ATA: AW and Spitfire at Hamble, 1942

14 Torcross summer holiday, 1947: AW with Vivien and Liz

15 Supper at the Mynd, 1946. *l–r:* Theo Testar, Pat Pringle; opposite Pat his brother John; Ray Wijewardine is in the shadows and Walter Morison with his back to the camera

16 Faithful VJ tied down safely on the hilltop at Salzgitter, Germany, 1946

17 The red Minimoa after Silver C distance to Lüthorst from Salzgitter

18 The Surrey Club Weihe coming in to land at Redhill, 1947

19 At Redhill with Lorne and the prototype Olympia, 1947

20 Surrey Club skiers at Zermatt, 1948. *l–r:* Hugo Trotter, Brenda Horsfield, Paul Blanchard and AW

21 A Swiss Mosway above mountains near Samedan, 1948

22 Jan, 22 years before qualifying as a doctor

and all components were present. The rocks and boulders were right on the edge of the steep mountain face up which rags of orographic cloud were racing to be sucked into the storm above. The valley far below was invisible. I tried to imagine Kit soaring this difficult mountainside and could only conclude that new misty cloud had formed beneath him to swirl up into the cloud above, with the Gull soaring only a few feet off the face. It would have needed only a few seconds of continued turn in cloud for Kit to be suddenly faced with this towering mass of boulders. He would have had only one course open, to pull up and hope to clear the crest which he would have known was not far above him. But in pulling up the Gull must have stalled, to drop square on the rocks. That was all that seemed to fit the pattern of the wreckage. There was nothing more I could do, so pocketing a small identifying piece of plywood I joined the other two sheltering behind rocks from the wind that was now blowing fiercely.

Our guide immediately set off like a horse turned for home, but across the mountain, not down. We followed, stumbling over loose stones in the half dark as the rain began, until without warning he disappeared into a small hole in the rocks. Inside, we found the calm of a shepherds' hut, warm with a huddle of goats. Our guide, still smiling, motioned us to sit on a narrow shelf running along one side of this cave, took a wooden bowl from a hook, milked a goat, and handed me a bowlful to drink. I had never been keen on either goat's milk or cheese, but this was nectar. When the fury of the storm lessened our guide took us back to his village, sat us down in a hut, shook hands and left us. He wanted no reward. Then an old man and his wife appeared, having a furious, almost pantomime, argument about whether he should or should not wear a coat. He won. The coat was left behind and he would guide us down to our car. He kissed his angry wife and we set off, reaching the Land Rover an hour later, and in the next village found an inn where we could spend the night. As soon as we got there I telephoned Samedan to report and to find news of Donald.

The police had discovered him that morning on the other side of the same mountain. He had flown into an unseen and unmarked cable used for transporting logs. This had severed one wing and he had been immediately killed on impact. Lorne and Peter Brooks had driven down during the day to try to find out more, and cope with some of the sad things which had to be done. Now, it was almost midnight and I fell on my bed exhausted and slept until first light, when we set off round the mountain to link up with Lorne and Peter and formally identify Donald, godfather to my small daughter Elizabeth. Flattened by events we returned together to Samedan, to a subdued prizegiving and a service in the English church arranged for us by Pirat.

DAISY 1949

8 · Redhill – into high gear

1948 was still close enough to the memory of war to yield feelings of hope and optimism. There was so much to do and look forward to, in spite of our shattering experience of Samedan, and we were young and ambitious for our gliding. The BGA, now with Dudley Hiscox as chairman, was full of ideas for the future, but the Ministry of Civil Aviation was also finding its feet, and wanted to bring gliding into its overall legislative and licensing system in anticipation of the great civil aviation boom that was expected to be just around the corner. Not unexpectedly we, in the BGA, wanted to continue to stay free to be responsible for ourselves, as the Association had been with increasing effectiveness since 1930. The threat was a real one and in February I was delegated by the BGA Council to press our case. Today, this sort of thing would be looked after – and needs to be – by the most powerful group that could be found, but in the early postwar years the BGA was still very small. So it was just me; and about the only advantage I had was that I believed absolutely that the only way that gliding would flourish was if we were allowed to look after it, and care for it, ourselves. As it happened I had the opportunity to meet Lord Nathan, then the Minister, and he invited me to discuss the BGA proposals with him in his office. Conscious of the need to succeed at this key moment I asked Lorne to come with me for moral support.

When we arrived Lord Nathan said that he had asked his adviser to be present to hear our case for self-regulation; so, taking a deep breath, I tried to say why we did not want state legislation, nor could afford it, and why I believed that the BGA was capable of looking after gliding both responsibly and safely. I almost certainly talked too much as I was terrified of missing out some vital point, but when I finally came to a halt Lord Nathan turned to his adviser and asked for objections to our proposals. Maybe the adviser was as keyed up as I was but without the advantage of having a cause, or maybe he had not done his homework, but whatever it was he had difficulty in marshalling enough objections to make a very good case.

'Well,' said Lord Nathan, when the adviser, too, had dried up, 'since there do not seem to be any strong objections I see no need to impose legislation. The BGA can do the work itself.'

We went back to Redhill in a daze, not yet able to believe the truth. At the time gliding, held together by little more than its enthusiasm, was so fragile that the Ministry proposals could have set it back years. There were still no more than fourteen clubs with only eight of them in a position to train new pilots. But with the threat of state control out of the way the BGA could concentrate all the efforts of its Council on helping the clubs. It should be for ever thankful to Lord Nathan for his forthright encouragement.

Soon after this Fred Slingsby produced his new two-seater: the large and slow T21 with side-by-side seats in an open cockpit. One of the first was coming to Redhill so we could soon be rid for ever of solo training; but teaching dual needed a higher level of pilot skill in the instructor as well as different techniques, so the need now was to find or make such teachers. As chairman of the BGA Instructors Panel I had to do something practical in this direction as well as just having plenty of ideas, so as soon as the Imperial College Gliding Club, who had joined us at Redhill, took delivery of the T21 – Daisy – we started both teaching and compiling a manual from the experience gained. We were also astonished at how much more quickly members were able to progress compared to the old solo ground hops.

By the spring of 1949 clubs were beginning to flourish all over the country; most of Elliott's first hundred Olympias were flying, the London Club at Dunstable finally had its clubhouse returned from duty as a POW camp, and by the time the National Championships came round in August the Surrey Club was able to enter three gliders with mostly Redhill-trained pilots.

The Derby and Lancs Club were to be the organisers, and although good at encouraging spectators to come up into the Peak District hills they were surprised when 10,000 arrived on the first day, in spite of petrol rationing; but also delighted to cover the cost of the meeting so soon. Not unexpectedly, Philip Wills again won the competition, but the Surrey Club Weihe flown by John Neilan, Wally Kahn, Derek Reid and Don Brown came second and won the Best Team trophy, while the Imperial College Olympia flown by Frank Irving and others came eleventh in their first contest. We returned home well content.

These successes were no doubt helped by our Redhill policy of encouraging cross-country flying by sending pilots off on their 50 km Silver C distance as soon as they were ready. There was also no shortage of retrieve crews among the newer members as every distance flight was still something of an adventure, with great excitement at the club when no telephone call from a pilot meant that he might have gone a long way. On 5 April 1950 Lorne had set off in the Weihe, borne away on a fresh northwesterly with a good chance of reaching Dover; but when his call finally came through it was not from the white cliffs, but from Brussels. After eleven years the Channel had been crossed again; though this time Lorne had managed to penetrate the sea breeze on the French coast to find lift from inland thermals. The problem now was how to get him back again. As I had retrieved Geoffrey Stephenson from the first crossing twelve years before it seemed appropriate to go again, so next morning VJ and I went off together for the 200-mile trip, to arrive at the main Brussels airport during the afternoon, and look down at the red Weihe sitting outside a hangar near an enormous new Lockheed Constellation. Next morning I flew the Weihe and Lorne the tug on the rough and windy $2\frac{3}{4}$-hour tow back to Lympne. Half-way across the Channel cloud thickened and lowered and in the misty air VJ was suddenly turned down Channel losing height so rapidly that I had a hard job following even with the airbrakes fully open – if the tow rope had back-released it would have been a cold dip. I wondered if VJ's little Cirrus had given up but the propeller was still turning. There was almost certainly nothing wrong but one of the most helpless feelings in flying is being on the end of a tow-rope and not knowing what the tug pilot is going to do next. A few minutes later the wanderings ceased, we turned back, and soon landed at Lympne. Lorne said he was trying to avoid towing me into suddenly lowering cloud, which, with the worse view from the Auster, looked a far greater risk than it did from my superior view on the end of the string!

The only shadow that fell on the happy Redhill days was that before too long they must come to an end. The RAFVR and the expanding flying club were bulging into our half of the airfield, and our growing activities were bulging into theirs, including the added problem of our winch cables. Not unnaturally everyone understood the hazard and was careful, but this did not prevent a Tiger Moth touching a cable one day resulting in the demise of the Tiger, though without serious injury to its occupants. The time for a change, sadly, had come.

After a great deal of searching we discovered that Lasham airfield, 40 miles away in Hampshire, was disused except by the Army Gliding Club, and that we could go there in August. The move was the biggest club expedition so far and inevitably was carried out in a spirit of enterprising chaos. One Olympia was aerotowed there and its trailer followed full of furniture. Jack Karran soared another one to its new home, and everything else travelled on a shuttle service for the rest of the week. Flying restarted the next weekend. But any big move produces problems. Some members found the greater distance from home more than they could manage, and it took time to attract new local members. As well as this, Lorne decided that he must exchange instructing for a job with more long-term prospects. But the Club had £79 in the bank, plenty of members happy to put in a great deal of work; and that winter we had our great Christmas party as usual, complete with pantomime produced by Hugo Trotter, Frank Kinder and Brenda Horsfield. It ended only when the generator failed at one o'clock in the morning through lack of fuel, and all the lights went out. Later, for a short while, the club's electricity problem was solved when a member found a cable sticking out of the ground of this old wartime airfield that was still rich in volts.

In spite of problems the club thrived on creating a home of its own at Lasham. The Imperial College had moved with us and we soon came to an arrangement with the Army Club for joint operations. The old control tower became the club house, was quickly fitted with a bar, and freshly painted by Dick Watson, and many others who possessed different artistic ideas. Huts were turned into bunkhouses, albeit cold ones in winter, and members brought a cooker and a sink so that we could have a kitchen. It was only after Hugo had discovered chlorophyll and given us green spaghetti and green eggs for breakfast when he was duty cook, that our new Committee decided to avoid any unexpected long-term effects by advertising for a more traditional caterer.

MADRID 1952

9 · We win at last

The 1950 international contests, now officially termed FAI World Gliding Championships, were held in Orebro, Sweden. They were well organised and so, we reckoned, was the British team – though with its sense of humour rather nearer the surface. Philip, naturally, was chief pilot, backed up by Lorne and Jock Forbes, all three on Weihes, with Pete Mallett on a Gull IV. I was again team manager and Pirat Gehriger was, this time, flying for Switzerland. Altogether there were twenty-nine competitors from eleven countries, including for the first time the USA, with a young Paul MacCready flying yet another Weihe. But we did even less well than we had at Samedan; in the final placing Jock was best at fifteenth, with Philip only twenty-seventh. There had been more excuse for Pete Mallett because his canopy had come off in the air leaving him in a high drag draught. It took two days for a new one to come from Yorkshire and while waiting he had flown with the perspex glued together like a jigsaw puzzle.

The winner was a local Swede, Billy Nillson, who seemed able to stay airborne longer than anyone else with, as runner-up, Paul MacCready, who flew faster than anyone else. Philip correctly continued to say that our poor showing was due to our pilots having insufficient serious practice,

We win at last

but there were other reasons in a competition which had reverted to distance flying as its main objective. There were many long – and cold – overnight retrieves, and whereas the Swedes and others had large warm vans with beds for the pilots we could afford only a miscellaneous collection of private cars, including John Sowery's 1926 open Bentley tourer which was excellent at preventing sleep. The geographical layout of Sweden with its hundred-mile-long lakes was also a problem for pilots without radio as it was easy to land half-way down one side of the water only to discover that the trailer was half-way down the other. We also had a further retrieve problem when the young Nick Goodhart enthusiastically drove Philip's trailer fast over a humpback bridge with interesting inertial effects on both the trailer and Weihe wings. They crept into Orebro at 3 a.m. to be met by me and an incredible local carpenter the Swedes had found for us. He looked at the broken wings and immediately started working at great speed, measuring by eye. By midday the Weihe was ready for a test flight, but although Philip later took off on the task it was too late to gain many points.

At Orebro competitors were billeted in a school with an international dormitory for the girls. That morning as I was getting up at 2.30 in order to meet the broken trailer Pirat Gehriger's wife, going to bed after a long retrieve, was muttering, 'I hate this country, it rains all day and the sun shines all night.' But it was Pirat himself who put Swedish solemnity to the test, incensed by the licencing laws of the time which required customers to order food if they wanted a drink and led to ridiculous plates of papier mâché peas which had to be paid for. One evening we had just finished a real meal when Pirat said 'Let's show these Swedes how to have a party,' and led competitors down to the narrow bridge which was the only entry to the main street. Car drivers coming into the town stopped, amazed at the sight of happy foreigners singing on their bridge. Then the police arrived but they had no chance. Whenever one started to speak he was shaken by the hand and patted on the shoulder with great friendliness, so they soon gave up, leaning against the parapet with hands to head saying 'Eet is impossible, eet is impossible.' Then they went away and our party continued until they returned with the janitor of our dormitory school. He put up his hands. 'Pliz to go back to school,' he said with such a sad expression that, still laughing, we did so.

I think the police really did not know what to make of us all as only a few days earlier I had been cycling at speed down the main street hill with Kitty Wills on the pillion when we were brought to an abrupt halt by a policeman who drew his sword to bar our way. 'It is forbidden for two womens to be on a bicycle,' he pronounced. 'It is only permitted for two childrens.'

The following Nationals were again to be at Camphill. Philip's influence could be felt in the decision to encourage individual entries so as to give each pilot more flying, and to set closed-circuit tasks with their emphasis on speed; but in the event only ten of the thirty-five entries were individuals and the weather was poor. But there were hopeful signs now of new contenders for the team; such as Nick and Tony Goodhart, BEA pilot Frank Foster, David Ince and SAS Colonel Tony Deane-Drummond. Even better, Slingsby's new 18-m span Sky (short for Slingsby, Kirbymoorside, Yorks) had taken the top two places. Philip began to feel happier. Before winter the British team for Spain in 1952 was chosen, with Jock Forbes, Frank Foster, Geoffrey Stephenson, Lorne, and Philip; all to fly Skys, lent by the ever-generous Fred. Basil Meads would provide Kemsley Trust money, and, at last, our gliders and retrieve cars (Vanguards lent by Standards) would have Pye radios. Maybe this time we would do better.

But nothing is perfect. In Madrid we had a strong team, good equipment and confidence, but the organisers lacked several of these things, and from the start the championships had a certain music-hall quality. Most of us were billeted in hotels in the city with the main road bridge to the airfield under major repair, the regulations were incomplete and in Spanish, and the telephones did not work. For two days we had meetings in an extremely hot room in four languages to try to sort out details of take-off procedure while outside pilots were trying to fly. But even this was not always easy as at the best thermal time of the day the tug pilots disappeared for their siesta. Our food was supplied to us on the dusty airfield by a truck which stopped, smartly immersing itself in its own dust cloud, and from which soldiers threw paper bags of food – a lunch and supper each. They contained a tepid omelette or piece of meat, dry bread and a huge tomato. For lunch it was just acceptable but with no cool storage positively bug-ridden by supper; so team funds had to be used to subsidise a more nourishing evening meal, usually in Madrid's Swiss restaurant, the Siete Picos.

This time there were forty-two pilots in the single-seat class and eighteen in the two-seat to represent eighteen countries at the line-up for the opening ceremony. The evening before we had each been given our national flag to hoist during the fanfare. All were different sizes, and ours only had the red, white and blue on one side, the other being plain blue. This clearly would not do so I telephoned the British Embassy. They were most apologetic but they had no spare flag. So I asked around for information on British people living in Madrid and found a small business in a back street; they had a flag and would lend it to us. To avoid anything else going wrong we did a trial hoist that evening on the metal flag pole, only to find that half-hitching the halyard around the pole to keep the flag up put it out of my reach; and since I had no intention of hanging on to it throughout the interminable speeches, we spliced in an extra length of rope. It was a wise move.

It was always a relief to have done with the opening ceremonies and to start competition flying. We had everything serviceable and polished, and already our radios were proving their worth. The first task was Free Distance, at which Philip excelled. But he did not not win that day; little Gerard Pierre of France, in his first world contest, flew 186 miles, and Pirat 183. Philip made only 126 miles, Lorne 116 and Geoffrey 60. But worse still, Frank Foster wrote off his Sky, and together with Helli Lasch of South Africa and the Swiss Schachenmann, was out of the championships. They had all flown into a shallow low-level gale blowing out from a line of thunderstorms which swung the wind through 180°

as they were landing. Several other gliders were also damaged in the same area but could be repaired. It was tough for Frank but depressing for the rest of us as, after all our preparations, we seemed to be doing no better than at Samedan or Orebro.

After a rest day had been given for the many repairs, the next task to be set was Pilot Selected Goal, with a 30 per cent bonus for reaching it; so it was important to get there. As usual the day started calm and cloudless as we drove out from the stifling city to rig our bright coloured gliders in the hot hazy air, wait for the first small cumulus to pattern the sky, and for launching to start. Pierre declared as his goal Leon, 180 miles to the north-west, while Philip and Geoffrey opted for Zaragoza to the north-east. By late afternoon all three had succeeded, which was excellent because with such a big goal bonus, those who failed to arrive would lose heavily. At the end of the day Geoffrey had risen from thirty-sixth place to tenth, Philip from seventeenth to second, while Pirat dropped to seventh. Only Pierre kept his first place.

After another unwanted rest day the weather was again excellent with high cloudbase and strong thermals, and a Goal Race to Torresaviñan was set 124 km distant. This time Philip beat Pierre, but not by quite enough to make him first overall. Jock stayed in fourth place, while Geoffrey and Lorne again moved up. We began to feel more cheerful, but unknown to us the organisers were far less happy; with the magnificent weather and the large entry they were running short of money for tow plane fuel.

At briefing on 8 July yet another rest day was declared, but competitors had not come across the world to sit in the sun on a dusty airfield and look at superb cumulus from the ground. We wanted no more rest and said so. The organisers gave in, declared Free Distance and launching started. Then someone overrode the decision and towing was stopped, resulting in a posse of team managers complaining that a task could not be stopped once started: and what about the pilots who had already taken off? They would send three planes after them to wave at the pilots to come back, they said; but the thermals were so good that the aeroplanes never caught up with the gliders! Some stern words were spoken as we had so far had only three tasks in six days. Next day Free Distance was set in weather that was no longer so good, but Philip did well enough to overtake Pierre, and Jock, Geoffrey and Lorne all continued their upward progress.

By now the airfield which had started with a covering of brown grass was mere brown dust, and the line abreast take-offs were becoming exciting in the fog kicked up by the first tugs to go. Most gliders, like our Redhill Weihe, still had undercarriages which had to be dropped after take-off, and collecting them as they bounced about in the take-off fog was even more exciting. It was about now that the telephone system reached its nadir, with some pilots taking more than a day to get through to report their landing place. Sometimes no sound was heard, but in a way this was less frustrating than 'Here is Gehriger', click, and nothing more. Our radios were envied. The Americans had an added retrieve problem as they had hired local vehicles. These were called Unimogs and looked a little like dumper trucks, but with the ride being made even less comfortable by the mechanic who had fitted the towbars. He had bolted them on to both the axle *and* the chassis, cunningly eliminating whatever springing the vehicles had to begin with.

On the final day the organisers surprisingly asked the competitors to choose the task – Pilot Selected Goal or Race. Inevitably the voting was equal so they had to decide anyway. They chose to repeat the race to Torresaviñan, but did not select the enormous thunderstorm that brewed up over the course. It was a matter of through it, round it, or under it; over it was not possible. Lorne went round while Philip and Jock went through, climbing to well over 20,000 ft. Philip's barograph ceased working near the top, depriving him of a height record, and Jock's oxygen failed, temporarily depriving him of his sanity. We had to censor the radio log.

This day was a triumph for the Americans, in spite of the Unimogs: Dick Johnson made the fastest time of 66.8 mph (107.5 km/h) in his homebuilt RJ 5 while Paul MacCready was second at 55 mph; but it was not enough to win. So

We win at last

Philip Wills became champion. We had made it at last.

The leisurely pace of this championships continued with one and a half days for packing up before the final banquet. This was a splendid feast of food and wine laid out on a 50-m long table on the airfield. Unfortunately the citizens of Madrid found their way to it faster than the now relaxed competitors. To have anything to eat and drink it was necessary to penetrate a huge, noisy crowd and pass back along our 'thin red line' anything we could lay our hands on before it was seized by someone else. Nevertheless it was a very good party and at 3 a.m. we all set off back to England with Philip's great big shiny pot.

10 · No dull moments

From the very beginning of gliding the challenge for the pilot has been with the elements, to use his skill to exploit the air's elusive energy, or to defeat its anger. It is a duel which the pilot tries to win by himself; single minded and on his own wings. So most pilots prefer to fly solo and most gliders are single-seaters; and there is no doubt that soaring solo over unfamiliar country is one of the most satisfying of delights. Even in championships most pilots want to compete on their own. Nevertheless, from time to time, it is fun to fly with another pilot of like mind, if only because in gliding it happens so seldom.

Before we left Redhill Hugh Kendall won a two-seater design competition and had managed to get his K1 built. Known affectionately as the Crabpot, it was a large and fast wood glider with a butterfly tail, and Lorne and I flew it together because Hugh wanted a second opinion on the terminal velocity with airbrakes open, and also on the spinning.

Our first need was to ensure that the vertical dives were in fact vertical, so we put a line on the left side of the canopy which would be parallel to the horizon when the Crabpot was truly vertical. While Lorne concentrated on screaming earthwards I would shout – it was noisy at high speeds –

'Steeper, steeper' until we were there, when I noted the airspeed reading. We soon discovered that the airbrakes did not limit the speed to anything remotely like a safe value, so we pulled out of the dive with some care. Hugh made some adjustments and we tried again. This time the brakes did limit the speed but played a new game by going irreversibly over top dead centre. So we now descended at a more civilised speed but at just as fast a rate and it quickly became apparent that the airfield could not be reached. The only good patch was a recently cleared woodland with tree stumps still installed. Lorne did his best to find a landing path between them, but as there was a little sharp one ahead I lifted my bottom off the seat as far as the straps would allow as we touched down. It impaled the poor Crabpot just behind me, so we spent the afternoon derigging and organising a quick repair. Hugh's next adjustment was just right. The spin was also interesting because there was a sort of airless pause at the stall sometimes followed immediately by a violent tuck until the Crabpot was almost on its back. Recovery ensued, but with a real risk of once again exceeding the maximum permitted speed before normality could be restored. It was a pity but the Crabpot could not go into use until these problems were overcome; which meant, even more sadly, that it probably would not be ready for the 1954 World Championships in Britain.

After these excitements Lorne and I were invited to what we expected to be a quiet championships in Yugoslavia, at Borovo. It was actually our honeymoon, as my previous marriage had not survived the combination of five years of separation followed by living with relatives during the immediate post-war housing crisis. It was a problem all too common at the time.

We arrived at Borovo after two days in the train just as briefing had finished for the first contest task. The June morning was warm and cloudless, and chickens were pecking among the parked gliders. We were briefly welcomed by Boris Cijan and Lorne was told, 'Here is your Weihe – it is new; here is your crewman, Marian Knes; and here is your towplane and pilot.' We could not believe what we heard but there it was before us. The tug was a Russian-designed Po–2 and it would be flown to wherever Lorne landed by Anton Shimek, who looked like Hollywood's idea of an angelic Mexican bandit. Boris continued, 'Here are your maps. The task is a race to Belgrade and take-off is in twenty minutes. Oh, your old friend Pirat Gehriger is here too.'

And so it all happened. Lorne took off in his new shiny Weihe in twenty minutes and was one of the eight pilots to reach the goal, although he fell asleep twice on the flight there and once on the tow home.

In the early 1950s Yugoslavian gliding was at a high point, being a strange mixture of unhampered enterprise and the great technical expertise of designer Boris Cijan. Since the Second World War, in a country beset by economic and other troubles, he had produced three very advanced gliders: the Orao, Kosava two-seater, and the all-metal Meteor (which was good enough to still compete in World Championships fifteen years later). His factory had also redesigned the Weihe to be stiffer, have powerful airbrakes instead of spoilers and a bubble canopy for the cockpit; it was one of these that Lorne was now flying.

Very late that day we finally got to the house in which we were to stay. We had been given the best bedroom with lace bedspread and curtains, and the most remarkable coffee for breakfast – half solid Turkish grounds, and the other half solid boiled milk, so the liquid content was negligible.

Borovo was marvellous. The weather was perfect and every day Lorne flew off somewhere in the Weihe, and I went with Anton in the Po-2 to retrieve him, flying the Weihe home on tow, while Lorne relaxed in the aeroplane. On one of these trips we landed in a small field near the Hungarian border to find Lorne and a pilot called Dimitrovski eating strawberries under the wing of one of the gliders. We took off as usual with me in our Weihe, and almost immediately met with another on tow: it was Pirat. This opportunity was too much for our three tug pilots who immediately moved into close formation with Pirat in the middle: the only problem being that the Po-2 wingspan was little more than half that of the Weihe, so we were jostling along in a most cosy fashion on the end of our strings; but the air was smooth, and the flat, flat land stretching away to the horizon beautiful in the late afternoon light. Suddenly Pirat's Weihe, on my left, started to behave in a most erratic manner, shooting up and then plunging below us. Dimitrovski, on the far side of Pirat, and I moved out away from him as far as we could, which was not much, and watched what might happen next. I could see Pirat's face clearly as he glanced in my direction from time to time but the weird dance of his Weihe continued, until an hour later when we reached Borovo and Pirat immediately released, diving down with airbrakes fully open. Dimitrovski and I also released but stayed high, floating in silky air until, with the purple shadow of night gathering on the eastern horizon, we sank reluctantly to the ground to land near the hangar. There was still a crowd round Pirat's Weihe and sounds of laughter. It appeared that shortly after take-off, when lighting a cigarette the box of matches had caught fire, and still burning had fallen under his plywood seat. Most of the Weihe's antics were due to his efforts to reach the conflagration as, apart from not wanting a burnt bottom, he was concerned for his parachute; so he kept wriggling this into a new position. His glances at me were because he reckoned that if his fuselage was on fire I would react in some way – he was certainly right there! His concern was real, as suddenly his harness came loose, charred off at the roots. Finally, school physics came to his aid: if he could cut off the air supply to the fire it would go out; so with exploratory care he felt with his fingers for every orifice beneath the seat and plugged it with the dried prunes he was carrying to ward off in-flight starvation. Finally the fire died but not because of the prunes – it just burnt its way through the bottom of the glider and fell out.

The fact that Pirat had prunes was simply because only prunes (and the strawberries which had been eaten) were easy to get and it was vital to carry something in the way of sustenance, as otherwise we would have starved. It was not that the Yugoslavs didn't feed us – when we had a meal it was enormous and very good – but the problem word was 'when', as the Yugoslavs seemed to be able to go for days without eating. Our crew, Marian Knes, could last for two days and we called him 'Camel', but such long periods

No dull moments

without meals caused a deal of rumbling in Swiss and British stomachs.

One day the competitors soared away to the south towards Skopje and by early evening only three pilots were unaccounted for: Lorne and two Yugoslavs; so three tugs were sent quickly to a small airfield some 150 miles away so that they could tow the gliders home early next morning. And early meant early, as clocks were the same as the sun – in midsummer it would be dark by 7 p.m., but sunny again by 3 a.m. So into the back of our Po-2 I climbed and at last light the three tugs were landed on a hilly grass field. From there we walked 2 km down to the town's 'Aero Club', a small shop in the high street, where we had a meal. Then, as the others departed for home, I began to wonder what would happen next. One of the locals, with no English just as I had no Serbo-Croat, indicated that I should follow him, so I did until we came to the entrance of a dark side-turning on the edge of town. There he indicated that I should wait, waved goodbye, and walked away. I waited a little nervously, a feeling that increased when a man appeared and, beckoning to me, set off down the dark little street. I followed, but several paces behind and ready to run; but I need not have worried. Shortly we came to the house of his elderly parents, who welcomed me in. For the next hour we sat at a table smiling at each other, and eating preserved plums with small pieces of ham. Then they showed me to their own bedroom freshly prepared for me. I really felt embarrassed; not only had I nothing for these kind people, but it was now midnight and by 2.30 a.m. I had to be on my way again.

At first light they called for me and told with signs that they had found an automobile to take me up to the airfield. At that moment an elderly car appeared, being pushed. It continued to fail to start and finally disappeared slowly round the street corner. We walked.

It was an idyllic dawn to be flying, clear and cloudless and absolutely smooth, and I sat in the open back seat of the Po-2 looking idly at the little villages that passed below. Eventually we came to a field near a substantial old town with three gliders parked in a corner, and a little low flying identified soldiers guarding them. Anton twisted himself round in the front cockpit and shouted 'Now we find ze pilots,' and dived for the town. I have never before, or since, been flown at 4 a.m. up and down the main street of any town at first-floor level until heads came out of the windows; but it worked, and three of the heads we recognised. We landed in the field with the gliders to wait for the pilots, who arrived on foot quite quickly, Lorne carrying a large meat pie he had bought from a street stall to ward off our usual daytime starvation.

The Po-2 was a distinctive aeroplane: a little bigger than a Tiger Moth it had a 125-hp radial engine which had five cylinders, three to one exhaust pipe and two to another, so it was instantly recognisable by its strange pop-popping noise. One day I was invited to fly it myself, and climbed into the cockpit to be faced with instruments in the Russian alphabet, indistinguishable from each other. Briefing was simple. Standing by the cockpit my mentor inclined his hand upwards like an aeroplane climbing. 'Hondred kilometres,' he said. The hand levelled off: 'Hondred kilometres.' Then the hand inclined downwards for return to earth: 'Hondred kilometres.' Clearly an aeroplane of sparkling performance. But I was still worried by one thing: I didn't know how much petrol there was in the tank, or even if there was any at all. So I said 'Petrol? Essence? Gasoline?' hoping that one of these would ring a bell. It did. 'Enoff,' he said, and he was quite right.

Borovo was a delightful holiday with Lorne doing well to come fourth in the contest, but it was good to get back to Lasham to find what had been going on; and to hear more about Slingsby's new two-seater Eagle, which it was hoped would be ready for the 1954 World Championships. This was to be held for the first time in Britain, at Camphill. There had been some opposition to using a hill site with only winch launching, and no aerotows, but Philip was really keen to show what British gliding could do, and the Derby and Lancs Club had the expertise. But the large entry was more than the site could cope with, so it was reduced to forty-four gliders from nineteen countries; though this was still too many if both Open and Two-Seater classes were to be able to soar in ridge lift at the same time to wait for the thermals. The weather was awful, and too often it was safe to fly only one class in the windy slot between cloud and hill top. The final count was five minimal contest days out of a possible fourteen.

This time I was not to be Team Manager, as Lorne and I had been selected to fly in the two-seater class with the new Slingsby Eagle. Unfortunately the cut-away at the wing centre section, intended to improve the rear seat view, produced instead some devastating drag. The week previously Lorne and I had broken the British two-seat distance record in it flying 154 miles from Lasham to Great Yarmouth, but strong thermals had largely obscured the poor performance. At Camphill, in the rough feeble lift, we could not maintain anything like the glide ratio of the Schweizer 2–25, let alone of the beautiful Yugoslav Kosava in the capable hands of Komac and Rain, who had been first and third at Borovo.

Our small two-seater class did not fly at all until the second day, when the eight of us were thrown into the air, to be bounced around in ridge turbulence under a grey overcast, which was so horrible that when a small blue break in the clouds turned up we all set off. It was too soon. Hans Nietlispach in the Swiss Spyr soared with us over a little hill near Chatsworth House, and had to land there while we managed to scrape along to Newark 40 miles distant; and only 5 miles short of the Kosava. But Nietlispach had been clever. His crew retrieved him quickly and got back to Camphill just in time to launch him into another short blue break in the middle of the afternoon, from which he flew 80 miles to win the day. But at least we had no broken gliders. The single-seaters, launched later, had a 73.4-mile (117.8-km) downwind race to Boston; two failed to get away from Camphill, four arrived at the goal, and five were damaged in field landings. There was, though, plenty of time in which to repair them, as there was no more competition for a week while it blew or rained and the mud grew wellie high. On one of these days the two-seaters were given a 40-mile race, but there was neither thermals nor wind. Lorne struggled back and forth along the ridge trying to stay airborne in case a clearance came, while, in the rear seat, I wrote letters to the children. Two days later the two-seaters were given Free Distance but by the time the short briefing was finished the wind was gusting to 50 knots, and the only thing to fly that day was a trailer. On 30 July the overcast cracked enough to permit a few afternoon cumulus and some wave lift.

Again the two-seaters were given priority as we had so far had only one contest day. Free Distance was declared, and was a triumph for the Yugoslavs. Rain and Komac recognised the weak lift they found as wave, exploited it and used five more waves, climbing to 6,500 ft above take-off and above cloud. In due course they saw Cranwell below them, found thermals, and landed at Marham 106 miles away. The next best were the Americans, Smith and Kidder, in the Schweizer 2–25 with 38 miles, followed by us with 25. Because cloudbase was always so low and thermals weak, it had been necessary to use any lift found, even at only a few hundred feet, and so that Lorne could concentrate entirely on staying up it was left to me to always keep a selected landing field in reach. In Derbyshire no field was large, and the surrounding stone walls were hard, so as we drifted slowly over the ground I would inspect each field as it came along, and be ready to say which one to go for when Lorne said, at about 300 ft, 'I need a field now.' On this day we were doing our usual low drifting act, circling in no-sink, and slowly losing out, and I had found quite a good field which was almost underneath when Lorne demanded our landing place. 'On the left, almost underneath,' I said. Lorne looked down and had swung into the approach when I realised that he was aiming for a dreadful little sloping patch next to the 'good' field, not having looked far enough underneath. My immediate thought was to tell him it was the wrong field, but knew that last-minute indecision would add to his problem, so stayed silent. Only when Lorne had safely fitted the Eagle into the fieldlet and we had stopped did he say anything. Actually we both spoke at once. 'This is a rotten field', and 'You've used the wrong field.'

Of the single-seaters Philip did best that day, contacting no less than nine waves and making 82 miles to Sleaford. Gerard Pierre, in second place, heard of this over the radio

No dull moments

while scraping up and down the Camphill ridge and decided to hang on. It was certainly a wise decision, because at eight o'clock in the evening weak wave developed within reach of the hill and Pierre climbed in it to dash off downwind. He landed his Breguet 901 at sunset 17 miles away, gaining just enough points to win him the Championships.

Then the weather worsened again and on 1 August only the eight two-seaters could fly. No one went far enough to make it a scoring day, and the Americans wrote off their 2–25 on a stone wall. So back around the world went the Argentines, South Africans, Canadians, and everyone else with little except muddy boots, new friends, and a hope that the sun would shine for the next championships in 1956.

As can be imagined these – and future – competitions were endlessly discussed. In future, poor weather – which could not be forecast – must be as far as possible circumvented by other factors, such as a lower latitude, a flat airfield with more space between the ground and cloudbase, and above all with aerotowing. The winch, even a row of winches, was too slow and too erratic to give competitors a fair chance. But Damphill, as it has since been known, was a happy championship. The organisation was good and in some strange way the challenge and the difficulties brought people together.

By the mid-1950s British gliding was fighting fit and full of confidence. It was not yet too expensive for young pilots and as they spread their wings on British Skylarks and Olys they pushed healthily into competitive flying. Lasham was up front, with a waiting list for membership and ready to take on anything that offered. So when the Derby and Lancs Club, after their 1954 World marathon, wanted a break from organising National Championships, it was natural that they should come to Lasham with its large airfield, aerotowing, and plenty of members willing to work. They stayed for fourteen years, almost without a break, growing from an amiable contest between forty slow gliders to tough competition between more than eighty high-speed ships. They changed, too, from the league system which I introduced to give our few top pilots good competition with each other, to the Open and Standard Classes of World Championships, bulging with talent. To begin with Philip won, then the names of Tony Deane-Drummond, the Goodhart brothers, Peter Scott and George Burton, one by one, rose to the top. For the first seven years the weather was superb, with the Lasham Anticyclone arriving on time each year to give everyone hours of fine flying and sunburnt noses, but then it deserted us and did not return. It was about the same time as our favourite Met Man, 'Wally' Wallington, emigrated to Australia, so he probably took it with him.

The Lasham Nationals were fun because so many members volunteered to use their summer holiday to help to run them, making the nine days a big party for both themselves and the several hundred competitors and crews, and they were good for gliding's image, particularly when Prince Philip came by helicopter to open them. This naturally brought thousands of spectators and the need for vigilance; as with launching just about to start we discovered a dear old couple having a picnic in the middle of the airfield, where they thought it would be nice and quiet with a good view of the flying! Shortly afterwards seventy aerotows thundered past where they had been sitting.

As I was responsible for setting the tasks each day I lived in our little home-built caravan on the airfield, getting up early in the morning to talk over the weather with Wally the Met so as to decide what sort of task flight would give pilots the best challenge. Sometimes it was obvious what to select, and at other times extremely difficult, but always an interesting exercise for one's personal computer. It was a matter of feeding in weather, average glider and pilot performance, as well as the competition objectives, and trying to produce a task which not only achieved those objectives but was enjoyable to fly. The biggest imponderable was, of course, the weather; it was all too easy to waste a crucial day in a major championship by setting the wrong task. But it was satisfying when it came out right.

Although the festival of the Nationals was important, these big competitions were only a small part of the busy world that Lasham had become in the five years since the move from Redhill. So much was going on that to the

No dull moments 113

several hundred of us who turned up regularly it was a way of life. Most had caravans on the grass strip between the perimeter track and the line of stately beech trees, and if possible arrived, together with children, on the Friday evening for the whole weekend. There was not only flying; Lasham was our home, and members came to build trailers, dig improved drains, or build a playground for the kids. Training frequently started before breakfast – a little bit depending on late retrieves or the party of the evening before. Sometimes it was made cheaper to fly early so as to encourage people to get up, and on Sundays we had a ballot at dawn for cross-country flying on club gliders. Privately owned aircraft were increasing in number, with more tug aeroplanes bought to launch them; and various other activities, such as parascending by the Scouts, and land yachting, went on in less-used parts of the airfield. As the years went by Dan Air brought in big jets for maintenance, and all this activity flourished without any need for formal air traffic control; even when Lasham movements were up to 50,000 a year.

One day a unit from the Territorial Army arrived on a corner of the airfield for some Exercise, and asked if we could bomb them to add reality! They gave me a 5-lb bag of flour. I took a young student for a free ride in the T21, gave him the bag with instructions to hold it over the side as I dived, and simply drop it when I said 'Go'. It was most embarrassing. The 5 lb of flour went straight through the open door of their radio van to burst inside with remarkable effect.

Much of my time at Lasham was, I hope, more constructively spent instructing in the kindly but draughty T21. Derek Piggott – who ran everywhere – was now full-time CFI, following a short period after the move from Redhill when Lorne and I, and then Paul Blanchard, had looked after training. It was fine having Derek there energetically in charge as it left me more time both for Instructor Panel work, and teaching individual students: one of the big problems in most clubs was that each beginner was often taught by too many different instructors. With most of them flying in their spare time this was hard to avoid, but it was usually not helpful to the beginner. As with any other instructor I also had to teach a succession of different students, but tried whenever possible to retain continuity with at least some of them by arranging to come when they could next get to Lasham. Another problem in such a busy club was that on some days, particularly if the thermals were good, there were not enough instructors, so one could be trapped into doing perhaps thirty flights a day. I liked this even less as it did not give time to find out enough about a student's background and previous flying so as to gear the lesson to his needs. I also liked to have time to talk to the student after the flight without feeling rushed, to make sure that he had obtained the maximum value from the lesson. To obtain time I would try to arrange with another instructor to alternate with me on the two-seater; if possible a new instructor, because with a little switching around of the flying list he could have easier students or check flights to help him gain the experience he wanted.

The reason it was essential to devote plenty of time and thought to each student is that flying solo gliders demands continuous correct decisions for success – more so than in an engined aeroplane – with no one to help the pilot decide what best to do. He needed to rely on himself and have confidence in his ability to do so: and the developed judgement and skills to produce this self-reliance and confidence could not come unless the pilot fully understood what he was trying to do, and was able to translate this understanding into instinctive handling of his aircraft. If his teaching was superficial, or he had misunderstood something, he would be less able to cope with even moderately difficult situations, such as his first few field landings – unless good luck was trying quite hard. So I needed time – maybe I was sometimes slow with a student – but I knew, if only from my own experience, that one learns much more from making mistakes than doing things right. I needed time to allow the student to experiment and get it wrong, appropriate to his stage of learning, so that he would be able to properly appreciate the hazards – without at the same time either frightening him or risking the aircraft. When teaching stalling, for example, it was much more important to give plenty

No dull moments

of flying very close to the stall so the student would learn what it felt and sounded like – and the ease with which he could stall inadvertently, than to 'do' stalls as just an exercise in the syllabus. Turns were another lesson on which I liked to give time since accurate circling is necessary for successful cross-country soaring. Until the new pilot understood how to control the turn well he would spend more time falling out of thermals than staying in them. Sometimes I had students who had been progressed too rapidly, and although now solo still had difficulty making reasonable turns, and were bothered by not being able to fly better. Helping new pilots overcome these sort of problems, and see them later become good pilots was specially satisfying.

Sometimes though, despair came close. I had one pilot, Chris, who had already been solo, who was passed to me as apparently incapable of judging a safe glide approach into the airfield. Every time he strayed too far and the instructor had to take over to get the aircraft back safely. There was no future in this, because without the instructor there anything might happen. On the first circuit I flew with him he undershot and I had to take control. After talking the problem through on the ground his second approach was slightly better, but I still had to take over to get back into the airfield. There seemed only one solution: to tell this student that on the next circuit I was going to do absolutely nothing to correct his errors, and if we undershot we would end up in a field. Not only would he have to pay for the trailer to retrieve him, but he would lose friends because the two-seater would be out of use for the rest of the afternoon; a threat indeed. I made it quite clear, and as he took off sat back to await events. My heart sank as he floated happily along only marginally in reach of the airfield and then left his turn too late. My hands were itching to take over so I sat on them. Eventually, the glider was turned into land – three fields short of the airfield – and flown hopelessly towards it as intervening trees grew tall. Eventually it was take-over or disaster. 'OK,' I said, 'I've got her,' and shoved the nose down, airbrakes full out, and with all the sideslip I could induce. Rapidly we sank short of the trees and made it into a small sloping field. I said nothing, hoping that realisation would sink in: it probably did because he was sent solo again soon after by another instructor who wrote in his log-book 'Approach good'.

During these years of gliding's expansion there was a constant stream of new instructors coming up for their rating tests, and these would be done by myself and other BGA Panel members at the candidate's own club or at Lasham. Again I liked to have plenty of time to talk with the new instructor, as well as carry out the formal technical and flying test; as before the days of Instructor Courses they often turned up believing that their job was to produce the maximum amount of indigestible information. Obviously the instructor had to know a lot, but it was sometimes difficult to get across that information was, in fact, something to be handed out sparingly. Not only should the student be given every opportunity – and help – to work things out for himself, but to have time to absorb new information and build it on to knowledge he already possessed. It was usually better to have such philosophical discussions walking round the 3-mile perimeter track, as it produced fewer distractions than sitting in the noisy clubhouse. It did not take long for news of this strange habit to spread around and instructors soon arrived for their test ostentatiously wearing comfortable long-distance footwear.

Nevertheless, the Panel work was beginning to show results, and the 31,000 two-seater instructional flights of 1958/9 produced only one damaged glider. But to go forward there is always something to be done and the problem now was no longer basic training, but the solo pilot who, having obtained his C certificate, found himself on his own. Without better follow-up instruction his first field landing could still be a hit-or-miss affair, reliant for success on a mixture of his original learning and native quick thinking.

The difficulties in overcoming this hole in the learning curve were considerable, mainly because club instructors were fully occupied with new members coming in to the clubs, but also because of a shortage of high-performance two-seaters and instructors with the soaring ability to use them effectively. Inevitably, too, there was tradition crying out that in the past pilots had managed to teach themselves to soar across country, so why was it now necessary for a club to have to buy a costly two-seater in order to teach them? But with the new, heavier and faster gliders being bought by relatively inexperienced private owners it was necessary.

In the meantime, outside in the bigger world of aeroplanes, there was still a lingering suspicion that glider pilots were not 'real' pilots; that flying a glider needed less skill to fly than an aeroplane. What we wanted was recognition so that gliding hours could properly be used towards airborne time for a Pilot Licence. I believed that any competent glider pilot could fly a light aeroplane with no more than a good briefing on handling and performance differences, so proceeded to put it to the test. Frank Irving and Paul Minton, who had flown only gliders, were willing to be the guinea pigs, and the agent for Tipsy Nipper lent me one of his little short-winged aeroplanes at White Waltham. Frank and Paul took their PPL medicals so as to be legal.

I first flew the Nipper solo to decide how to do the briefing, and gave a DIY talk to Frank and Paul, sitting them in the cockpit to familiarise themselves with the controls and view. Then we went to lunch. Afterwards I gave Frank a quick re-cap briefing and off he went solo to make an excellent circuit and landing. Paul did just as well, so now we had genuine evidence that glider pilots knew how to fly just as well as anyone else. For myself I was sure that a good glider pilot probably knew more about the techniques of flying than some aeroplane pilots.

During these years No. 1 Test Group had the new family of Slingsby Skylarks to assess. Most of the testing was done by Lorne and Frank Irving, though every now and again I would be hauled out of the school two-seater to confirm some handling characteristic – or more usually to be given my old friend c.g. aft spinning, as I was still the lightest pilot. On the Skylark 2 the c.g. forward tests had produced no problem, and neither had there been any on the calculated aft limit, but the schedule required that the spin should be investigated still further aft. Before doing this test I had of course spun the Skylark within limits, so I was familiar with its normal recovery behaviour and, like any other test pilot,

I had worked through possible problems in my mind – usually in my bath.

When the right sort of fine weather arrived with the good visibility needed to count the required five rotations against some landmark, I took off on an 8,000-ft tow. The usual Lasham scene lay below in the clear air; car tows from the main runway with closely packed gliders waiting to take off, and the separate aerotow launch point with its own colourful queue. As we climbed higher so the circuiting gliders became little bright ants following each other on some timeless trail. Releasing in the high cool air I watched the friendly tug dive away and I was on my own, looking round, checking the kneepad pencil was still in place, and tightening my harness. Easing gently back on the stick I slowed the Skylark, softening the sound, and as the lift died brought it full back and put on rudder. Away she went, fast. I counted the 3-second rotations using a distant white house while watching for any tendency for the spin to alter or flatten, but it stayed steeply nose down. Four turns and down to 3300 ft; five – good, that's it. On with full opposite rudder, and stick steadily forward until the spinning stops. The white house flashed past again – six, seven – 2700 ft, and we were spinning as merrily as ever. I was now holding full rudder with the stick against the instrument panel, trying to decide whether to slack or undo my harness so that I could use my weight to get the centre of gravity forward, or whether to bale out; but I knew all the horror stories of pilots being pinned in the cockpit or over the wing unable to get clear. I decided to release the harness and get forward if the Skylark had not made it out by eight. The white house shot past, the sun broke through the overcast, and at the same instant the spinning just stopped. There was no shuddering or warning and once more the world was normal. Carefully I pulled out of the steep dive, checked position and height – just on 2000 ft – and the little Skylark was flying as nicely as ever. It was good to float around in the sunshine with the test safely done and sink towards the little coloured ants still busy on their trail. Soon I was low enough to join them, but overshot so as to land outside the hangar where Lorne and Frank were waiting.

No dull moments 119

'Maybe we overdid the sums a bit!' Lorne said as I handed him the kneepad.

Next day I spun the Skylark again to check a less extreme aft c.g. and she came out immediately on request.

Another glider we had for test was the T49 Capstan two-seater, in which I had a special interest for a training aircraft as it possessed characteristics nearer the single-seater Skylarks than the T21. It was also side-by-side; a good feature, as the instructor could watch where the student looked, and how he held the controls. And it had superb all-round visibility so there was no excuse for not learning to look out. Unfortunately the prototype suffered from interesting stall inconsistencies, so it seemed a good idea to tuft the wing (with Woolworth readicut rug wool) and photograph what happened from another aircraft. It was the Royal Navy who came to our aid. They could get a good training flight out of this for one of their helicopter pilots. So we aerotowed to Lee-on-Solent and later took off, with the helicopter managing to keep its downwash clear of the Capstan while still flying close enough to take useful pictures. Lorne proceeded to do a series of stalls while I watched from a reasonable distance in the tug. All went well and after thanks all round at Lee we flew home. Next day we heard the worst. The helicopter pilot certainly had a worthwhile training flight but the camera had not focused properly. The Capstan stall problem was overcome finally by increasing the washout, which meant altering the jigs to ensure that the twist remained linear on this straight high-taper wing.

Some of the most frustrating, but often funny, testing we had concerned glider performance. For this we needed not only the calm of dawn's first light; and not all of these were satisfactory, but to know the precise airspeed at which we were flying. This meant calibrating the airspeed indicator using a trailing static, which consisted of 50 ft of rubber tubing with a bomb-shaped object on the end obviously designed to penetrate anything it hit. For take-off this thing hung against the outside of the cockpit with the rubber tube leading in through the clear vision panel, and neatly coiled by one's right knee – or that was the theory. In practice bouncing over the rough ground on take-off caused total uncoil so that by the time one was fully airborne the cockpit resembled a zoo reptile house. Unravelling this lot and getting it out through the tiny window was an acquired art. Coming into land was, usually, much simpler. When, at about 60 ft and over grass, the dangling tube was released from the special cockpit instrument and writhed its way to the ground, hopefully missing some unsuspecting new member who had not yet learnt to expect anything, however strange, to fall from the sky in his new sport. Sometimes the air turned out not to be as calm as expected, so the flight was used for other tests, such as aerobatics. Lorne once looped the Gull IV forgetting he still had the trailing static on its 50-ft rubber tube, the pitot bomb giving a wild scream as it was flung into orbit round the Gull.

But all that was in summer. Winter weekends were also spent at Lasham except that the first weeks of the New Year were still for skiing. With Jan, the youngest, now three and old enough to start, Liz and I took her to Davos where we stayed in a very small hotel with the same friends each year. As soon as she was accepted by the small children's ski school Liz and I would beat it up the mountains and set off down one of the long runs from the Weisfluhjoch. By going fast we would have a fantastic run and be down in time – just – to collect Jan from school. Skiing in the high mountains in cold, crystal air, especially off the crowded pistes where we could sail along with a gentle rustling sound through the unmarked snow was blissful. In a strange way it was like soaring in a glider, a sort of quiet exploration. This was quite illogical, of course, because thousands of people had already passed – or flown – the same way, but when there was no one else there you could almost feel that you were the first. I suppose there was a little of feeling oneself against a potentially hostile environment, but the immediate delight was that it was all so beautiful to look at.

But it was always good to get back to Lasham.

11 · Trailer days – and nights

The two years until the next World Championships always seemed an age away; until suddenly there was barely time to prepare. The 1956 Worlds were to be in the Massif Central of France, at St Yan, and once again as team manager I had to get everybody and everything together. The BGA had chosen Nick Goodhart to fly the reshaped Eagle in the Two-Seater Class with Frank Foster, plus Philip and Geoffrey on Skylark 3s in the Open Class. We arrived at St Yan in high spirits. The weather was good and the organisation looked good as well, with everyone living on the airfield in large army tents which smelt of North Africa, and provided with quantities of excellent food and wine. In an even more cheerful mood, if that was possible, were the pilots and crews from Poland, Czechoslovakia and Hungary who had been allowed out for the first time. A contingent came from Turkey, also for the first time, in resplendent gold braid, and including the only girl pilot. On the first practice day the rudder fell off her rented glider and fluttered down into a nearby field as she climbed away on aerotow. After release she could not understand why, whenever she joined another glider in a thermal, it promptly fled. Finally she landed, the glider ground-looping in spectacular fashion though without damage, and was astonished when someone arrived carrying her undamaged rudder.

Once we were through the usual tedium of the opening ceremony the flying at this Championships was impressive, although the entire British Team again succeeded in doing badly on the first day. There were two tasks at St Yan in which the challenge was probably greater than in any previous contest; both of them on days with a 50-knot mistral wind blowing down the Rhône valley away to the south. This created every opportunity – thermal, ridge and wave lift – and every problem, with low cloud, wild turbulence, and difficult navigation. The second competition day was one of these, with the British Team naturally determined to retrieve its reputation. The task given at morning briefing was Distance along a Line through Cuers, not far short of Toulon on the Mediterranean coast. It was, however, lunchtime before the thermals brewed, while further south near Lyons they lost heart and were feeble. Those pilots who survived this patch reached the mountains with good lift over the ridges which lay across the route, and from there they found wave. Philip climbed to 15,000 ft, and although Nick and Frank in the Eagle achieved only 8000 ft and 211 miles distance it was enough to take them from bottom place to top in their Class; and really raise our morale. By early evening only a few Open Class pilots were still airborne, flying high with superb views of the Alps, pink in the evening light; among them were Philip, the American Bill Ivans, and Saradic in the great Yugoslav Meteor.

It was nearly dark when they landed at Cuers, congratulating each other on best for the day; but none saw Paul MacCready float silently overhead in the dark to land at a lighted military strip 9 miles further on. The trailer crews had left at the same time as the gliders for a possible 800 km of driving, though ours had, as usual, departed before the rush leaving myself, Ken Owen from *Flight* magazine, and Wally the Met to look after launching our gliders. It was a plan that worked well, the trailers pausing at some suitable radio vantage point until sure that no pilot was going to need a quick retrieve from near the airfield. Since Spain in 1952 we had improved both our radio equipment and techniques and now could not only deal effectively with retrieves, but from base quickly produce weather data or even assist a mislaid pilot to locate himself using our own set of maps.

The great battle with the mistral wind came again on the very last day. By now Paul MacCready was unassailably leading the Open Class by more than 1000 points, but Nick and Frank, having fought their way up to top place in the Eagle, could still lose to the Yugoslav Kosava. Well before dawn it was obvious what the task would be, as the wind was blowing so hard that pilots and crews were out in pyjamas tying down the trailers. At briefing it was certain. The task was again down the Rhône valley, this time a race to the gliding school at St Auban, 188 miles away. The surface wind was already 40 knots, and over 70 at flying heights. There would be people out on the airfield at St Auban, they said, to hold down the gliders after landing. The real problem with this wind as it roared through the mountains would be the wild turbulence; for some pilots more than they had ever experienced before. But this was a World Championships where, to win, fear and caution had to be harshly controlled. Philip, with his massive experience, did not have this extra burden to carry and soon found that the best way to cross the great cloud-covered ridges across his path was to rise in the hill lift, heading straight into wind, and allow himself to be drawn up into cloud until his altimeter told him he was above the peaks; then to turn downwind, breaking through the overcast into the gloom of the next valley and, steely nerved, repeat the process. At 10,000 ft he reached St Auban; as did Geoffrey, but many others ended up in little sloping meadows among the mountains, and Bill Ivans crashed on a remote slope. The only pilot to see him was Pelle Persson, winner at Samedan eight years before, who had already landed and ran 2 miles across the rocks and ridges to help him. Only seven pilots reached St Auban, among them, not unexpectedly, Paul MacCready.

At prize-giving next day, which finished abruptly in a violent thunderstorm, we were delighted when Nick and Frank, champions in the Two-Seater Class, received their medals. It was a historic moment because this class would now disappear in favour of a much-discussed and cheaper

Standard Class, for which Lorne was helping to draft the rules. We were happy as we drove home, stopping in Rouen and eating together in a restaurant charmingly called Grill Jeanne d'Arc. We had won at two Championships out of five – and with British gliders.

There is not space in these pages to go into the background of World Championships; the endless discussions on rules or how the team should be selected; the finding of sponsors and lenders of expensive cars or radios; the spares, the accounts, the thank-you letters – and the planning once more for two years ahead. At home we were increasingly using the National Championships as a proving ground for potential team pilots; which in turn brought criticisms from the Clubs that the BGA concentrated too much on competition and not enough on them. It must have seemed so because it was the Championships which attracted publicity, not the quiet, creative work of the Clubs, though in the long term the issue was simple: if we were going to enter World Championships at all we had to give of our best – the alternative was not to go, and this was unthinkable.

The 1958 Championships were to be behind the Iron Curtain for the first time, at Leszno, Poland, and the British Team would naturally include Philip Wills and Nick Goodhart, but Frank Foster had since been killed when a military jet was in collision with his BEA Viscount over Italy. Newcomers would be Tony Deane-Drummond, chosen for St Yan but left behind when the French limited the Open Class to two per country, and Nick Goodhart's brother, Tony. The difficult problem was to decide which two should fly in the Open Class and which in the new 15-m span Standard Class, so it was passed to the pilots to choose in order of selection. Nick and Tony Deane-Drummond opted for the Open Class, and would fly a Skylark 3 and Elliott's new Olympia 419 respectively. This left Tony Goodhart, who was happy to fly a smaller 15-m span glider, and Philip Wills who was not. His faith had always been in large spans, and the years had shaped his soaring technique to suit: but everyone knew he was a great person for rising to the occasion.

So we again collected the new loaned Vanguard cars and

Trailer days — and nights

made up an extra strong base team, because so little was known about Polish repair facilities, roads, telephones, and the bureaucratic problems with which we could be faced. In addition to Wally the Met, and aerodynamicist Frank Irving, who was crewing for Tony Goodhart, we added Harry Midwood to look after repairs, and John Williamson, son of my long-ago friend Henry, in charge of radios and as reserve pilot. Ken Owen came to work as well as write, just as he had done at St Yan.

Our great expedition started most pleasantly, being invited to tea at Buckingham Palace with Prince Philip, before we set off across the Channel on the night ferry for the drive east. We had hoped to cross Czechoslovakia in the day but found the going slower than expected, particularly as Tony Goodhart's trailer swung about wildly at speeds over 35 mph. So I went ahead of the trailers to Prague to get some money and arrange accommodation. This done I parked prominently in Wenceslas Square to await the trailers. There were very few cars in Czechoslovakia in 1958, with foreign cars almost unknown, so the Vanguard was quickly surrounded by an eager crowd, fingering its shiny paint. Yet this was nothing to their interest when I started transmitting on the radio, calling the trailers to give my position. Heads leaned in through the open car windows, goggle-eyed, and entranced at this public display of illegality by smiling foreigners: but it was perhaps just as well that I had contacted the trailers by the time that Czechoslovak voices, puzzled and angry, started breaking in on our transmissions. It was a time to be tactful and shut down the station. Even so, later that day we were shadowed for some miles by a big black Tatra car.

We arrived at Leszno the following afternoon, expecting to be met by some dour military organisation with drab uniforms and unsmiling faces. But it was not like that at all. The Championships Director was a woman barely 5 ft tall, Irene Zabiello, with husband Roman helping. They were cheerful and friendly, and gave us a marvellous book in delightful English setting out everything that they had arranged for us. On our tent accommodation it said 'For each crew there is one tent. If there are women among the crews they can according to their wish stay with their crew or in the special tent together with other women. If she wishes to stay with her manly crew the screen will be provided.' So we requested screens, but they fell down whenever anyone hung their clothes on them, and in any case their near transparency produced, as Nick's crew Bryan Jefferson remarked, only tantalising silhouettes. Since the gloom in the dark brown army tents was enough to protect any feelings of modesty, we forgot about the screens and requested duckboards and mirrors instead. The duckboards never arrived but that afternoon two enormous mirrors appeared, bought specially for us in the town.

'You will receive meal cards for three meals a day cooked in Polish style,' the book continued, 'and when the pilots will fly they additionally receive the dry provisions.' The quantity of food we were given was unbelievable – 8,000 to 9,000 calories a day – probably to show us that Poland was not starving. Fortunately some of this was packets of biscuits – 'Keks' – and chocolate, and we had no problem on retrieves in finding children only too happy to relieve us of any surplus. The pilots' flying food was not, in fact, dry, but consisted of tins of sardines and bottles of fizzy drink. One pilot opened his at 5000 ft spraying himself, the instruments, and the inside of the canopy with instant sticky orange.

'To the disposal of the members there are the showers.' I used them at 0545 each day so as to have them to myself, as any second user guaranteed the water temperature fluctuating through 20°C in as many seconds. One morning, just as I got it right, the entire tap assembly took off, shooting past me propelled by a powerful jet of water, which landed neatly on my clothes. Philip had a different encounter in the showers when just before take-off he had a quick one as he was convinced he had collected a flea. While still damp he dusted himself off with insect repellent – except that in haste he picked up a tin of Nescafé.

'Your personal linen can be given the quick laundry in the administrative tent.' This was a dark, cavernous marquee filled with cheerful local women with tubs and flat irons heated on braziers and was most efficient. The marvellous

Trailer days — and nights

briefing book continued after several more helpful pages with 'Understanding Form', which described a paper for pilots to give to local people after landing, with instructions in Polish to telephone Leszno, so all the foreign pilot had to do was smile and use the opportunity to hand out a few more 'Keks'.

We had several days' practice to get used to this flat, green, sea-level country, the language problem (for the first time), the unpredictable telephones and the weak beer: this was just as well as the water was suspect so there was little else to drink, except when persuaded to try the local vodka – there was a distillery in almost every village. The organisers, particularly Roman and Irene, wanted so much to make these Championships a success that nothing was too much trouble for them. I had to be careful not to ask for anything under the heading of luxury – the enormous mirrors provided for our tent were a constant reminder.

By the end of the practice week we had found both our level and the strength of the opposition. There was no doubt that the still formidable Meteor flown by Saradic, and Ernst-Gunther Haase's new German sandwich construction HKS looked more likely winners than our reliable and kindly Skylark 3 and Oly 419.

The three practice tasks were all triangles for speed: 106, 208 and, for the first time ever, 300 km. On this one our best, Nick, was 27 minutes slower than the Meteor. Philip started off on the triangle but he wanted so badly to gain his 500-km Diamond, for which he had tried many times, that he rounded the two turn points, so as to be observed, and then set off for the Tatra mountains in the south of Poland. He landed at last light but, once again, just a few kilometres short of the magic 500, returning to Leszno just in time for the opening ceremony.

This was a remarkable display, including a glider dangled from a helicopter and dropped at 600 ft to gain enough speed for two slow rolls before landing, followed by the release of 3000 pigeons for peace. Then came the anthems and the flag raising. What we did not know at the time was that the Japanese had come without their 'discwy of the national hymny'. When, with no time to spare, Irene had discovered this she seized the Japanese pilot – in Poland for his first competition – and set off for the Leszno town fire brigade whose band was practising for the ceremony. Faced with the extraordinary problem of creating an instant oriental anthem they responded without hesitation. Irene commanded the Japanese to sing and the band to listen. Then she turned to the band and commanded 'Render'. Apparently satisfied with what she had heard – or realising that she was on a lost cause – she dashed back to the airfield to get ready for the ceremony followed by the band. When the moment came we wondered why the Japanese team, with incredulous expressions on their faces, were a little hesitant in rising to their feet.

It was odd how national flags and anthems, which have been around for a long time, continued to produce new and unexpected incidents. At one FAI Conference many years later the awards ceremony was held in a very warm hall and continued long enough to induce sleep – including the conductor of the military band on the stage. Long before the final citation was read one of the dignitaries near him scraped back his chair. Instantly the conductor woke, leapt to his feet and raised his baton for the national anthem. Straggling somewhat, three hundred people rose mystified at this unusual national custom of playing the anthem in the middle of a ceremony.

Unfortunately the weather during the Leszno Championships did not live up to the promise of the practice period and the first five days produced only one task. Then it perked up and Free Distance was given in a west wind which could bring the Russian frontier within reach – and another 500-km opportunity for Philip. By the end of the day ten pilots had exceeded this coveted distance – including Nick who was the only British pilot to already have gained it. Once again poor Philip fell just short. The retrieve cars as usual travelled much further, as can be seen from this account by Frank Irving in *Lasham Magazine*, who was crewing for Tony Goodhart:

> Minor roads can be anything from good tarmac to loose stone and dust, often badly corrugated. An extreme case was

Trailer days — and nights

encountered by Nick's crew, who found the trailer aground at both ends with the wheels gently revolving in clear air in the middle. Driving a trailer on a road where the dust lay three inches deep was rather like driving on ice, and any tendency for the trailer to get unstable produced spectacular oscillations. We once waltzed round a corner in a highly non-geometric fashion to find ourselves in the midst of a flock of ducklings. They all escaped, by some miracle, but the clouds of dust, feathers and sinusoidal tyre-marks in our wake clearly impressed the locals.

However, leaving Tony's launch in the hands of the base crew, we set forth for Gostyn and a small rise in the flat countryside whence, so it was said, the car radio would reach the airfield. This turned out to be a fallacy, so we were faced with the telephoning situation rather earlier than anticipated. A calculation involving the latest time of launch, possible retrieving speeds and the weather, suggested that Pleszew was a suitable point. Telephoning in Poland is an operation to be planned well in advance and undertaken as infrequently as possible. Even armed with official 'understanding forms' the operation was likely to take up to $1\frac{1}{2}$ hours. There were, to use the celebrated phrase, a series of interrelated bottlenecks, both administrative and electrical, and the final achievement of coherent communication left the crew member on the verge of a nervous breakdown. Equally enervating was the task of guarding the car. The locals were pleasant, eager to talk and help, and very curious to observe the glossy western motor and its chaotic contents. After being the focus of interest to about 100 children and townsfolk, all breathing heavily through the windows and thrusting autograph books into his hand, the crew was usually exhausted.

However, Pleszew was known to contain a taxi-driver who spoke English, so we enlisted his aid and learnt after only half an hour that there was no news. The soaring conditions looked good, so it was clearly important to get to the Poznan-Warsaw road and start some fast driving. Needless to say, the next 20 miles involved frightful surfaces but once on The Road, we were able to make good time by pushing the speed up to 50 mph. In retrospect, it is difficult to visualise Tony's trailer travelling at such a speed but it is remarkable what can be done in the absence of police cars lying in wait. However, one had to remember that the reactions of cart-drivers, cyclists and pedestrians are attuned to a low traffic speed and the reactions of animals were utterly unpredictable. The day wore on, with beautiful cumulus clouds, and as a further telephone stop at Lowicz gave no message, we resolved to go through Warsaw and telephone again from Minsk-Mazowiecki or Siedlce.

Warsaw is dominated by the Russian-built 'Palace of Culture', a building of overpowering and singularly tasteless aspect. It can be seen rising from the plain from many miles outside the city, even before the trams and cobblestones start. East of Warsaw, we encountered Kitty Wills, in pursuit of Philip, and shared a telephone call (1½ hours). There being still no news, although now nearly 8 p.m., we carried on to Siedlce, where Kitty enlisted the services of the Police, the Post Office being closed. Notwithstanding their diligence, it was not until about 10.30 p.m. that we learnt that both Philip and Tony had landed close together about five miles from the border, towards the south-east, near a village called Zablocie which, inevitably, was not on our map. Relying on the policeman's pencil-mark, we shook off the remaining few dozen children from the trailer and departed into the darkness. Once off the main road, we settled down to driving over the sort of corrugated roads on which you go at 10 mph or 40, but not in between. Not only was it dark, but horse-drawn carts were creaking about in the gloom, the only indication of their presence being a minute mud-covered reflector hanging from the back axle. We settled for 10 mph. At about 1 a.m., we came to a T-junction, and were just investigating some singularly obscure sign-post when a policeman emerged from the darkness. After some exploratory remarks in sundry languages, it became apparent that we had half-a-dozen words of German in common, which had to serve. Slightly doubtfully, he announced that glider No. 55 was one way and No. 56 the other. Just as we were about to go our various ways, he added that both pilots were sleeping in the village, and we were to follow him. After a considerable detour to park his bicycle, we all marched off in his wake up a narrow lane straight towards Russia, causing one member of the party to remark that when we reached the River Bug we really would be up the creek.

However, a farmhouse eventually appeared and after the policeman had waved his torch through a window, the cultured tones of the Chairman of the BGA floated incongruously through the warm Polish night, gently wondering if Kitty had arrived. We were then treated to a silhouette of Philip putting on his trousers whilst expressing heartfelt gratitude to the good farmer. Tony was reputed to be still lurking in his field, so after expressions of mutual esteem on all sides, it was

Trailer days — and nights

back to the trailers, where the policeman pointed the way to No. 55, and said that the road to 56 was bad. As he shone his torch up a rutted river of mud, this was seen to be something of an understatement and the Wills crew spent the next three hours on an extensive tour of the local marshes before getting within striking distance of the Skylark.

We encountered Tony when a great cry came out of the darkness 5 km before we expected it. He was dozing in the cockpit with two farmhands keeping watch from a pile of hay in the ditch. He seemed to be in a remarkably small field, but it turned out that he had been pulled there after landing in an adjacent marsh, a few trees having been chopped down to facilitate the move. We de-rigged, presented English cigarettes all round, had a swig of canned beer and set off for Warsaw. Choosing a route which seemed more direct, the next 150 km were conducted at less than 15 mph, the trailer brake-rods fatigued through for the second time, and we arrived with only enough energy to order breakfast. Then to Leszno: 747 miles in 31½ hours.

Each day, after briefing held only in French and Polish, we needed to agree a precise translation and work out our strategy. The base crew and I would get the trailers away early, launch our gliders, and then keep watch on two separate lots of telephones – those that took incoming calls from the police about landings and those reserved for crews out of radio range. Only by listening for any English names to pop out of the flow of rapid Polish could we find out about our pilots and take any necessary action – or discover that all was well.

One day, before the trailers were safely away, the loudspeakers announced with great brevity, 'L'orage s'approche'; though it was already obvious that the thunderstorm approaching itself was coming our way, and fast. I put my car on the expected windward side of our precious stores tent and tied the two together, Kitty used Philip's car to protect their trailer the same way, and the other trailers I had hitched to their cars so that they could be driven on to the airfield and kept dead into wind. In minutes the wind rose to 50 knots and it became too dark to drink beer in the hangar. National flags, stiff as boards, caused their tall poles to bow low. The large number of two-foot-high letters which spelt 'World Championships' in Polish blew off the hangar to fly around in the heavy rain like alphabet soup, and a kiosk roof and two trailers took off for distant parts.

In their great efforts to provide for our entertainment, as well as for the flying, Irene announced on the rest day of 23 June that 'In this part of Poland there is every midsummer a great festival of fertility rites', describing it as a 'night of carefree love'. Today was the day and it would be celebrated at the castle of Osiecznej, where gliding's technical congress (OSTIV) was in progress. It sounded promising so as evening approached the world drove in its cars to the castle 'by the lake lovely situated'. We were sitting drinking on the terrace in the warm evening darkness looking at the floating bonfires and a long line of canoes with lanterns out on the lake, and listening to haunting Polish music when another of the special local brand of thunderstorms struck. Everyone fled for shelter, some of us climbing to the castle's high turret to watch the impressive lightning strikes and, with Pirat, inevitably talk about the future of championships. It was only many years later that Irene told me that she had invented this 'traditional' festival for our entertainment – but that the local people had enjoyed it so much that they had continued it ever since!

In spite of Nick's 500-km flight there was no chance of his catching Haase in his HKS with its thermal sensor secret weapon (which did not actually work), but he came second. Poland's own Adam Witek became Standard Class champion. Philip came thirteenth, only just beginning to get to grips with small-span techniques.

After prizegiving we spent a happy afternoon flying each other's gliders, then packed quickly. We wanted to leave now as soon as possible to avoid the sad change from the exciting, living entity that is a championship to the disintegrating nothingness that so quickly follows the end. One minute the whole place pulses with laughter and life and suddenly it is gone, and all that is left is a deserted airfield with bits of paper blowing in the wind. We were away from Leszno for the frontier at first light, to chat a bit on our radios while driving along the East German autobahn to keep the locals on their toes, crossing to England the next day.

On the strength of Haase's win with the HKS West Germany offered to hold the 1960 Championships at Butzweilerhof airfield, near Cologne, which at first sight appeared to be no more than a forest of radio masts. For the British team it was the usual mixture of endeavour and improvisation. We were well equipped with radio, cared for by John Willy – known as the Rundfunkfettlemeister; and though it was the rule for pilots to telephone Championship Control after any outlanding we naturally monitored everything ourselves. One day we heard on our base radio that Tony Goodhart had landed not far away in Cologne's Beethoven Park, and immediately directed his trailer there to rush him back for a further launch. While waiting he telephoned Control. Less than an hour later he had again landed in Beethoven Park, and again telephoned Control. 'We know this already,' they said, dropping the note in the wastebasket.

On 8 June the task was distance along a line to the north, but the weather turned out better than expected, and numerous gliders reached the Baltic coast to land on the beach; but World Championship pilots do not give up easily and out to sea there was the small island of Fehmarn. Without warning its lonely inhabitants found themselves as a honeypot with no less than nineteen enormous bees all looking for somewhere to land. Five went into the farthest field, and the only ferry shuttled trailers most of the night. But life is never simple for championship organisers; in the middle of all this Dick Schreder, USA, radioed that he was over the Baltic at 600 ft – he didn't explain why – so seven German navy vessels were dispatched to search. Dick did reach land, only to discover he was in East Germany and he wanted to come out again!

It was a good Championships, though for me it had the autumnal feeling of the end of an era. There was no obvious reason for this as the classic Skylark 3, flown by Argentinian Rolf Hossinger, beat the sleek new Polish Fokas and Zefirs with their pencil fuselages and supine pilots, and Nick finished fourth flying the old Oly 419; though if a day cancelled by the organisers because of a rules interpretation problem had counted he would have been champion, as he had won the task on that day. The real cause of coming change was at Butzweiler, but it was not so obvious – the German Phoenix, flown by Haase, was the first white glass-fibre glider. It was regarded with some doubt as an experiment, but it was to herald the end of the bright coloured wooden gliders – and many long retrieves. Few visualised that within ten years there would be more than 10,000 glass gliders, and the jump in performance they provided would revolutionise World Championships.

12 · A little flying on my own

With almost every week filled with instructing at Lasham, taking children to school, cooking, or having an occasional sortie in the garden with some destructive tool in a desperate attempt to avoid total jungle, I suddenly realised that it was some time since I had done any proper soaring on my own: worse, it was thirteen years now since completing my Silver C and I had achieved nothing towards the Gold C or Diamonds introduced later. Worse still, I was the one who complained about other instructors sitting in the two-seater all day, and becoming narrow in their outlook, and here was I doing just that.

From Lasham the popular way to fly the 300-km distance for the Gold C was to declare Perranporth on the north Cornish coast, using an anticyclonic north-east wind. In the dry air cloud base was usually high, and if a good speed could be maintained it was possible to arrive in Cornwall before the peninsula was cleared of cumulus by the afternoon sea-breezes surging in from both coasts. If Perranporth airfield was both declared and reached, the flight would not only count for Gold Distance but for the 300-km Diamond Goal. Since the West Country was for me like home, it would be a delightful flight to do, so I kept my maps with a line already drawn.

But my Gold turned up in France, not Cornwall. Earlier in the summer of 1959 Nick Goodhart had taken his Skylark 3 to Switzerland to do some mountain soaring. He arranged that Lorne and I would drive over there in my Bedford with the children for a camping holiday, collect his car and trailer, and do some flying ourselves with the Skylark. It had the makings of an excellent mix of business and pleasure and as soon as we arrived Lorne had some fine soaring in the high Alps above Samedan, while the children rode in the cable cars. Then the weather collapsed, curtains of cold rain hid the high snows, and, while wondering whether to stay and hope or start home, we met a gliding friend, René Compte, who advised us to get north as soon as possible. But Nancy in France was just as wet and windy, so after camping for the night at St Dizier we set off for the gliding club at La Ferté Allais just south of Paris under a clearing sky; Lorne towing the trailer with me following in the Bedford. It was a chance visit as we did not know whether we could get a launch there, or even if foreigners would be welcome, but the clouds were now beginning to look good, and it was probably this preoccupation with the heavens that caused us both to go the wrong way down a one-way street. The policeman looked at our strange convoy, at three pairs of children's eyes staring at him, and decided that it would be simpler for us to continue on our way than to reverse through the town – clearly a man of sound judgement. At 11.30 we arrived at La Ferté and went straight to the office of the Chief Instructor, M. Pechaud, to ask if it were possible to have a launch. He looked up from his desk, took in everything he needed to know at a single glance, and said in English. 'You want to fly distance?'

'Yes,' I said. 'Three hundred kilometres.'

'Then it is not too late if you go at once, *but at once*, and you must fly to Angoulême.'

Too astonished to speak, we rushed out of the office, I to tell the girls what was happening, and Lorne to find help in rigging the Skylark. He had no problem; already local pilots had the glider half out of the trailer, and in minutes it was ready, and Lorne had started to draw the magic line to Angoulême on his map for me. Thirty minutes after arriv-

A little flying on my own

ing I was airborne on tow. But now came a problem. Certainly I had a map – an enormous concertina of a map, as Lorne was going through a navigational phase of sticking all his maps together – but I realised suddenly that, having just followed another car for the last hundred miles, I did not actually know where I was starting from – unless I could find one end of the pencil line. Flying the glider while trying to unravel this heap of map resulted in sufficient meanders on the end of the tow rope to cause the tug pilot to look round sharply and increasingly often. But all I could see among the voluminous folds was Madrid and then Munich, which produced a distinct feeling that it was all getting out of hand, and it would be much simpler to land back and have a peaceful lunch in the sun. Then I suddenly found a bit of the line and at the same instant the tug pilot waved me to release.

Keeping my thumb pressed hard on the pencil line I looked around, spotted the airfield I had left, and saw that the cumulus were now impressive with bases at around 4000 ft. The thermals were not very strong, but plentiful, and I reckoned that if I could get up to cloud base I might be able to sort out my navigational confusions and then go on my way. To do the 300 km with the few hours permitted by the sun meant that no time could be spared wandering vaguely off track, so I needed at least to start in the right direction. Off tow and climbing, the monstrous map was beaten into submission and I found the La Ferté end of the line with the route to Orleans and beyond. At nearly 3000 ft I set off.

All went well for a while until, approaching the city of Orleans on the great bend of the Loire, cloud increased until above me was a grey overcast, with no sunlight on the ground anywhere ahead. No more sun meant no more thermals, and as I flew gently on in smooth subsiding air, the day seemed as though it was finished. 'Oh well,' I thought, 'that's the way it goes – worth a try though,' and I was just turning back to save Lorne driving any more of the interminable miles of the last few days when the sun gleamed momentarily on the featureless ground ahead. So I flew slowly on in the murky air, circling in any flicker of lift, over a land of lakes, clear on the map, but to the eye undistinguished grey puddles.

Crossing the Cher river raised my hopes as the dreary cloud began to break, with shafts of sunlight picking out the river bends with silvery flashes. Then suddenly I was in the clear. Cloud base rose rapidly to 5000 ft, so I set off to fly fast and use only strong lift; but this was almost my undoing because these powerful thermals ceased to be easy to use. In the new air they were rough and narrow; and they were far apart. I knew that I could not have much time left, but had no idea how much as I had broken my watch the previous day, so I tried to stay as high as possible; but kept being forced into long straight glides, sinking steadily into the hot air low down, each time with a hopeless feeling that this was the end. But whenever my concentration transferred itself from the clouds to the ground to look for a landing field, another rough, tough little thermal tipped me over and instinctively I swung into its lift. But I gained little from them, as they all soon pushed me out again to vanish for ever in the bright sky. Then I saw some bonfire smoke and went for it, to ascend slowly 2000 ft in its aromatic thermal to a dusty cloud, happy with this reprieve. Visibility had now become almost infinite and I had a fine view of Chateau Chauvigny before crossing yet another big glittering river, the Vienne. But the sun was now well on its downward slide so I had to pussyfoot along, trying to make sense of the difficult little thermals. In spite of their ill-temper I managed to stay airborne – luckily with the only cumulus still growing in the faded sky directly on my line. The light northerly wind was also gently drifting me on my way, and as I approached the Charente river I found myself peering over-optimistically far ahead for any signs of my goal.

Then came another long straight glide through air like silk, and this one went on and on until hasty calculations told me I would reach the ground almost exactly in the middle of a forest just this side of Angoulême. Slowly I flew on, praying for just one more thermal, however miserable – just enough to carry me along in air buoyant enough to cancel the Skylark's sink. Now I could just see the airfield in the distance and was willing myself to get there somehow. Then the air fidgeted just a little. 'Wait for it, don't turn too soon!'

The bubbly feel came again and this time stronger as I finger-tipped the Skylark gently round, holding my breath. Suddenly, roughly, we surged up into the strongest and biggest thermal of the day. It was glorious. The forest sank down into insignificance, with Angoulême airfield there for the taking. But I couldn't leave this beautiful thermal, fate might never be so kind again, so I rose with it to 4000 ft, the sun glittering across my instruments with every circle. Only when it died finally did I turn for the airfield, double checking that it was the one I was supposed to go to, and floated quietly high overhead in the pale light until gravity gently brought me in to land. It was exactly 6 p.m. The airfield was deserted except for one other pilot who looked at me doubtfully when I said I had come from La Ferté. 'So have I,' he said, a little firmly I thought. He had not taken off until eleven that morning and it seemed strange he had not seen me there. He was only reassured when I telephoned to say that I had landed.

To my astonishment, Lorne arrived at midnight with the trailer and children, having driven 400 miles that day. He had had to burgle the gates of the airfield to get in and was somewhat abashed when his shout of welcome was met by my 'Shut up you fool, you'll wake the guards.' At least that's what he always insists I said. But they stayed asleep and we had a funny few moments looking in through the uncurtained windows of the fully lighted guardroom at the North African soldiers fast asleep, before we moved into the empty airfield bar to put up our camp beds for the night – to be woken at first light by four-year-old Jan, who had been carried in asleep the night before, demanding, 'Mummy, what country are we in today?' But returning home through the British Customs was worse; when loud and clear it was 'Do I have to declare my Colorado beetles?'

Before I left Poland at the end of the 1958 Championships, Roman and Irene Zabiello asked me to come back, promising all sorts of wonderful things such as unlimited free flying, but it wasn't until 1961 that this became possible, when Lorne and I were invited to their National Championships. Lorne would compete, and I would be technically team manager, but with the main objective of trying to get my 500-km distance, for which I was lent a 17-m span Polish Jaskolka.

But not until the Championship was nearing its end did the weather become suitable for distance flying across the length of Poland to the Russian frontier. The day gave promise of high cloudbase and strong thermals but they did not grow early, so it was near eleven o'clock by the time Lorne, flying a Polish Mucha Standard, and the other competitors took off. I was launched shortly after the last one, and so would have to hurry to make the 500-km distance before the thermals died away in the late afternoon.

The Jaskolka had a glide ratio of 1:27 so was not a high-performance glider by today's standards, but it was good to fly and I was happy to be on my way over this land with its villages connected by straight, dark green lines of trees hiding the roads beneath. Navigating was easy, so throughout the hot midday and afternoon I sailed quietly along from one stately cumulus to the next, but anxious that to the east the sky appeared to be clearing early and the clouds dying – flying 500 km eastwards actually shortens the day by 20 minutes; and I reckoned I was going to need all the day there was. So I flew more slowly now, searching out every flicker of lift, and circling in it with a light touch, trying to stay high. But my last thermal had been still weaker, lifting me to only 3000 ft; a warning. The air quietened as I flew on, sinking gently. I had to accept that the day was finished and so was the chance of my 500-km distance – for which Philip, after all, had tried many times. The nearby cumulus were dissolving and towards the east, and Russia, the sky was clear down to the purple-tinged horizon. It was seven hours since I had taken off from Leszno and now in silky smoothness I had to look for a landing field. Ahead lay the meandering Vistula river which I reckoned I could just cross, though still 125 km short of the magic 500. Logic said there would be no more lift ahead but hope is not always logical, and as long as I was in the air something might come my way. Down to 1500 ft; just enough to get over the river, except that it seemed to be without bridges and through a country without roads. It would make more sense to land this side – a quicker retrieve, perhaps a hot bath. Why go on? There was no more lift anywhere; but I could not bring myself to turn back. Only 1000 ft now and in air totally calm, like a dream. 'Concentrate – you have to find a field.' I stared across the water, half determined to fly on until my wheel furrowed the ground while the other half of my mind demanded commonsense. 'You have only three hours of daylight left in this desolate country; and you have to find a telephone – and a bed for the night.' But closer now the fields across the river looked better cultivated. Still no houses visible and no long lines of poplars hiding roads. At 500 ft I crossed the swirling yellow water, examining the fields this side of a wood. It was then I saw that the trees were hiding houses, and that a big field this side of them contained some crop young enough to let me land between the rows; but still no people. Perhaps the hamlet was deserted; other glider pilots had done just as I was doing now in populous, wealthy America and landed by a ghost town. Instead of a friendly welcome there had been creepy silence, broken by a sighing wind and creaking doors straight from Hollywood. One pilot had spent the night in his cockpit after a supper of bilberries and the crumbling remains of his sandwich lunch. I pushed such thoughts away to concentrate on the landing, but this was easy, needing only a gentle turn in the smooth air to line up between the rows of seedlings. With a gritty rumble my Jaskolka slowed to a stop and for a moment I sat there quietly among millions of baby spinach plants.

The warm scented air was beautiful when I opened the canopy, undid my parachute and harness, tucked my maps away and looked at my watch – quarter past six. Carefully I wrapped up two 'Keks' and a half-eaten apple, hoping that they would not have to be my supper, climbed slowly out under the pale sky from which I had come, and looked around. Nothing. Then from the trees, running across the baby spinach, came two small boys. They arrived breathless, and stood there on tough bare feet, in cut-down trousers, staring at me. Instead of a young Polish man there was a foreign woman; not even a young one.

They looked at each other, and without a word turned

and ran. Then one stopped, came slowly back and stared again. I smiled, said 'Telefon' and 'Allo, allo' into an imaginary instrument. The boy grinned, unconscious of his broken front tooth, and nodded vigorously, pointing to a building in the trees. Soon the second boy returned accompanied by what seemed to be the entire village traipsing across the spinach quite regardless of the little plants. They arrived breathlessly chattering, the women smiling at me, the men not so sure, while the boy stood close, pleased with himself. I repeated my request for a 'telefon'. More chattering, then the broken-toothed boy pulled at my sleeve and we all set off for the trees. I was the only one to walk between the spinach plants. The boys took me to a small brick building. 'Telefon,' one said, and sure enough a wire ran from the roof to disappear across the fields; it was the only telephone in the village. Outside the building a small, dark-haired man was lazing in the last of the sun, his policeman's tunic open. He stayed where he was while the boys told him about the glider – I kept hearing the words 'szybowiec' (glider) and 'telefon'. The man looked up at me without smiling, got slowly to his feet, and put out his hand; but not to shake. I produced my Understanding Form containing the telephone number and request in Polish that the pilot should be helped. He read it very slowly, so slowly that I wondered if he might not do better with it upside down. I pointed to the telephone and said please – prosze was about my only other Polish word. He read the paper all over again. I was thirsty and tired, the usual effects of any long flight, but tried to contain my growing frustration. Finally the paper was folded and the man returned to his reclining posture on the grass. 'Niet,' he said. The two boys ran off and I was on my own, staring with disbelief at this stupid man. It was now after seven o'clock and I knew that if I did not telephone soon my retrieve would be delayed, I could miss my train reservation to England, but more infuriating, a quite unnecessary search for me might be started. I went over to the man who watched me with a closed-up expression, and again said 'Telefon, prosze,' tapping my watch and hoping to get across some sense of need. He shook his head. The situation was becoming ridiculous. Deciding to force the pace a little I went into the building, found the telephone, and picked up the receiver. He was behind me immediately and banged it back on its rest, saying 'Niet' angrily. I looked around then and it was apparent that this room must be the village police station – the obvious place for the only telephone. A further look revealed that there were only two rooms and the other one was the cell. As if feeling he now had the upper hand the man indicated that I could sit in the cell. It certainly had fewer mosquitos than outside so I sat on a chair inside and he sat himself on one outside to guard the telephone. Every few minutes I made further persuasive efforts, drawing sketches of the car and trailer which was to collect me, and a little map of Poland to indicate the 400 km it had to come. I made signs to show that I was hungry, which by now I really was, and I showed him on my watch that it was now nearly 8 p.m. It was no use and I could have cried, my imagination picturing me in this wretched village for months. The villagers had been friendly but had disappeared. I felt sure that it was only this man who was the trouble – perhaps this miserable telephone was the biggest thing in his life. Then I decided that I had had enough; time was becoming too critical. I would go and find the villagers and see if some other solution might not present itself. I got up, said 'Telefon, prosze,' once more, received the parrot response, and walked straight past him out of the building. He jumped to his feet too late to bar my way, but followed me.

Without looking at him I went towards where I hoped the centre of the village might be. It was; though little more than a dirt farmyard, and as I arrived so people appeared from every direction. They stood close about me, all ages, talking to each other. They seemed not only friendly but sympathetic; perhaps they did not like the policeman either. I held up the paper with the telephone number and tried by signs to get across that the policeman would not allow the telephone to be used. Maybe this had happened before because the chattering increased, hands started pointing in different directions, and the children ran off towards the fields. The little policeman stood on the edge of the crowd with no influence on events, which raised my spirits. Pressed by this crowd of poor people who appeared to have my interests at heart my feelings of despair began to subside. Then I felt a hand grip my arm and before I could snatch it away a quiet voice said, 'Parlez-vous français?' Quickly I grabbed the arm like the lifeline it was and looked at its owner. She was a small, worn, peasant woman, frail in stature. She smiled up at me. 'Don't worry,' she said in French, 'they have gone to find the Commandant. He', indicating the glowering little policeman, 'is only the under-policeman. The Commandant – the over-policeman – he is a better man.' She seemed pleased at the obvious relief which must have shown on my face and then cut another chunk off my worries by saying, 'If you wish you can stay in our house tonight.'

Before I could thank her the first of the children arrived back followed by a tall, pleasant-looking man in uniform; clearly the over-policeman. Immediately the world became a lovely place again. In the same breath and at the same time everyone brought him up to date on the story, and action immediately followed. He told the under-policeman to do up his tunic properly and at once go to the field of baby spinach to guard my glider; no interpreter was needed. He then led the way to the policehouse and got Leszno on the telephone. The French-speaking woman, still beside me, kept up a commentary, telling me it was too late to collect me tonight – which I knew – but they would do their best to arrive with the trailer sometime tomorrow afternoon. Then the Commandant asked if I would like to see if my glider was all right for the night. I said I would and was informed that 'An auto will come.' It was now quite dark but with a yellow moon, and the night was warm and scented with lime blossom. There was no wind so I knew the glider would be safe but as it was not mine I wanted to be sure. Then the auto came; it was a very elderly motorcycle and I was helped on to the pillion by the wives of both policemen, that of the under-policeman giving me a pullover for her husband in case he was cold in the night guarding my glider. She did not seem in the least upset. The over-policeman kick-started the great machine and with a final steadying push from the wives we were off. We did not go, as I expected, along any of the dirt

A little flying on my own

tracks I had seen, but along one I had not observed – a single-line railway track – and at a goodly speed over the sleepers. It was an extraordinary experience and I had to hang on to the over-policeman to stay on the bike at all. Then we bounded across the spinach and stopped by my glider looking beautiful in the moonlight with the under-policeman sitting on a kitchen chair in solitary splendour. I gave him the pullover, but could not help the sneaking feeling that on another occasion he might not be so reluctant over the telephone.

We returned along the railway tracks, making a noise like a frenzied geiger counter, and the Commandant set me down at the house of my new French-speaking friend. The house, in fact, was a very small flat above a newly built working men's club, and it was crowded with ornaments. Hot, beautiful food appeared and afterwards we had coffee. It sounds marvellous and it was, but it could have been the only coffee they had as they went on adding water until the last cupfulls were almost transparent. We talked far into the night in the sort of halting but comprehensible way that is possible when people speak together in a language not their own. All this time a young man with a microscope was working in a corner. He was the local doctor and this remote farming area was his first job; well after midnight they insisted on my having their best bedroom and I fell immediately asleep.

Late the following afternoon the retrieve crew arrived and after goodbyes all round – even the under-policeman shook my hand – we set off, finally arriving back at Leszno a day later. There was just enough time to turn in my glider and maps, pack for my return home the following morning, and hear about what other flights had been made – Lorne had missed his 500 by only 7 km and had already left for England. I slept, exhausted. Almost at once, it seemed, there was furious banging on the door.

'Get up, quickly.' It was Roman and Irene. 'Hurry, hurry.'
'What's the matter?'
'Get up now, the weather's going to be perfect for your 500 km. You must fly.'
I turned over, pulled up the blankets and shouted for

them to go away. I was not only tired but in a few hours my train left for home. The banging continued. 'You must take off by nine o'clock.' That was Irene's voice. At this moment no suggestion could have seemed less attractive. I shut my eyes and countered with the information that I had handed in my glider.

'They're getting another from the store,' Roman's voice this time.

'I've got no maps.'

'Josef is preparing maps! You have to go to Tysowsze. It's easy to find, just nine kilometres before the Russian frontier.' Was there no discouraging these people? I started to face up to the inevitable: I was simply not going to be allowed to go home on that comfortable train, dozing and looking at the view. If they had their way, and I had a deep down feeling they were going to, in eight hours time I could be back at the other end of Poland once more, battling with other under-policemen, and taking two more days with almost no sleep returning to Leszno in a bumpy retrieve car. Such thoughts were too daunting. The banging became urgent and louder. 'This time you'll get five hundred, the weather is better. Hurry, hurry.'

Reluctantly I got out of bed, looked sadly at my packed luggage, and tried to face up to another day of concentration and of probable failure – but, my inner self insisted, a chance not to be missed. Roman and Irene rushed me, eating a slice of bread, by car to a small paddock near the airfield, and there was another Jaskolka, storage dust still on its wings, but already attached to a tow plane, with the engine running. I was pushed down on to the bare wooden seat – no time for a cushion – did up my parachute and harness, tucked my map and a tomato down beside me, and was just starting my pre-flight checks when the tug pilot opened up and took off; there was no time to smile or wave to Irene even if I had wanted to. It was incredible that every important opportunity in my life seemed to need grabbing as it flew past. Other people could spend days preparing while I did not even have minutes. Then I was airborne and once again trying to sort out my map on tow.

Lift was weak after releasing and I dithered about, wasting time that I knew I did not have to spare, failing to make much of the early thermals. I worried about crossing the Leszno forest low down, and having to land on the far side too late to return for another try. Then I stopped worrying. I was committed anyway, and at just over 2000 ft I turned away eastwards, slowly gaining 500 ft in another feeble thermal. But soon the sky began to look better and within an hour of starting, cloud base was up to 3000 ft. By midday there were cumulus everywhere and most of the time I was up with them, selecting the next cloud to go for from the look of its shadow on the ground. Now I could fly faster, cruising between thermals at the fastest effective speed of 115 km/h. At exactly 3 p.m. I reached the 300 km mark, and with cloudbase just over 6000 ft, I could discard any lift under 3 m/s. This meant that I rarely needed to circle, simply slow up through any thermal and then accelerate away again to save every second. At 4 p.m. I crossed the 400-km line. Now just 125 km to my goal and I could get there if I made no silly mistakes.

Again I crossed the wide, wandering Vistula, this time at 5000 ft instead of 500, although this day, too, was beginning to weaken. To be safe I decided to now use any lift I came across better than 1 m/s, and to stay high. I could not bear the thought of landing just those 2–3 km short. It was a bleak and empty land below with only a few remote villages and I did not want to find myself a part of it. By 5 p.m. the cumulus were collapsing all around, yellow and ragged as they evaporated to nothingness, while I crept along, over-cautious perhaps, but still above 3000 ft.

Unexpectedly, I saw another Jaskolka quite close, and we joined each other, circling together with the sun on our wings; we must both be making for the same goal. This was a great relief as there had been no time for anyone to describe what Tysowsze was like – I didn't even know if it was an airfield or just another indistinguishable bit of this feature-less land. I looked east into Russia, which was just more green flatness to the grey horizon. Suddenly the other Jaskolka turned and dived away. He must have decided he could reach the goal so I watched until the sun glinted on his wings when he turned to land. So I now knew what the field was like, and it certainly was no airfield, but I could see a little group of cream-coloured gliders on the ground; perhaps other happy pilots with new 500-km Diamonds. I dived for the field, watching the other Jaskolka being pushed by dozens of children, and at 5.55 p.m. landed on some hard rough ground just a few feet from a weed-covered grey-green duck pond.

From now on it was all straight pantomime. After eight hours cramped into a wooden seat I had difficulty getting out, and when the Polish pilots in their kindness to this foreign visitor said that I could have one of the only two aerotows out of the field that night – 'the other pilots will sleep under their glider wings' – I knew that there was no way I could stand any return to that wooden seat. 'OK,' they said, 'when the Po-2 comes, you go in its back cockpit and another pilot will fly your Jaskolka; he will be happy as he will sleep in a bed and not under his wing.' Soon two Po-2s rumbled in from the now clear and golden sky, and I climbed into the back of one, even more glad that I was not in the Jaskolka as the tow rope was only 14 ft long! The Leszno 85-ft ropes always seemed short, but this looked impossible. Maybe the pilot in my Jaskolka had done it before, as he had no problem, and after a 20-minute flight we landed on Rzeszow aerodrome, to await the other Po-2 with its glider. It did not appear, so as it grew dark the Poles arranged for a minibus to take us into the town for a meal – none of us having eaten all day (the Poles were like the Yugoslavs in this respect – human camels). It was then we heard the Po-2 'iddy-umpty' sound. The airfield had no illumination so we shone the minibus lights into wind and waited for the Po-2, also without lights, to 'pop-pop' in over our heads. It landed without difficulty, followed a few minutes later by the glider. After a great deal of laughter someone told me that the Po-2 pilot had got lost and had flown around until it was dark when he knew a high radio mast would have its red light switched on. When this happened he flew to it and from there knew how to locate the airfield. I was even more delighted that I had not been sitting on my hard wooden seat with a 14-ft tow rope behind *that* tug.

A little flying on my own

Next morning early Irene sent a Yak-18 specially to fly me back to Leszno, to start my overdue journey home. The smiling pilot in the front seat flew at less than 100 m the whole way back, his harness undone and draped over my rear seat rudder pedals, and stubbing out cigarettes constantly on the no-smoking notice. But he did not go even a few metres off track, as I could check from my map. It was a remarkable performance.

Even though I now had my longed-for 500 km, which also brought with it the Women's British National Goal Flight record of 328 miles (still held in 1983), it was marvellous to be invited again to Poland the following year – just for a holiday with my friends at Leszno. I did some soaring but the weather singularly lacked good thermals, so Roman decided that he, Irene and I should be tourists instead. He had 12 hours' flying to do for some military reserve commitment and he could do it anywhere in Poland in the Gawron, a large and cumbersome but quite comfortable four-seater. So off we went, first stop Cracow and a cultural tour from air and ground, refuelling free at the military base, and then up into the Tatra mountains to visit their old gliding friend Adam Dziurzynski at his hilltop school at Zar. It was sad that there was a prohibition on taking photographs from the air as we too never flew higher than 100 m so there was plenty to look at. This height, it seemed, was normal for light aircraft, so I reckoned the risk of collision between aeroplanes and gliders in Poland was minimal, as cross-country gliders rarely flew lower than 3000 ft.

To reach Zar we flew up a big wide valley, south of Katowice, now filled with stepped lakes for a hydroelectric power scheme, and then into a side valley to land in a long up-sloping field knee-high in millions of wild flowers. As our wheels threshed through the blossoms on landing their scent filled the aeroplane, swamping the usual smell of oil and dope. We taxied through more flowers to a beautifully built little stone hut to be met by Adam, who, having greeted his old friends, insisted that I look at his beehives. He produced an enormous hat with netting hanging down all round and beckoned me to follow him behind the hut where there was an ominous buzzing. He didn't have a lot of

English but explained that each summer from this lovely field his bees gave him the same weight in honey as a two-seater glider, about 600 lb. The tour of inspection over, we set off to go 1000 ft up the rear of the mountain, standing on a flat car which was pulled on narrow rails by a winch at the top. Normally used to carry gliders back up the mountain, it had no sides, and as far as I could see no brakes so I prepared myself to leap off into the bushes at the first sign of any cable failure. Signalling to the winch driver at the top was simple but effective – just a matter of giving the cable a couple of whacks with a stick, when lo and behold the car started. As we rose up the shadowy slope more of the Tatra mountains appeared, the hills and peaks taking on the golden glow of evening, while our flowery field sank into a blue dusk. We reached the top near 9 p.m. I was starving and longing for the promised supper, but Adam was so proud to show a visitor his club that he crowned his welcome with an invitation to fly. My heart sank, we had been flying all day and I was hungry. I tried the excuse that it was late and all the helpers were wanting to put the last gliders away. 'No,' he said, 'they were happy to launch me with the rubber catapult rope.' As I looked out over the tree-covered ridges and the artificial lakes, I countered, 'But it is a heavy two-seater and there is no wind to help us get airborne.'

'No matter,' he said, 'we will use two bungies and sixteen men.' It was clearly not going to be possible to refuse further, but as I was not sure whether I could even find the blossom-filled field around the back of the mountain in the dusk, I put in my last plea: 'But you will please come with me.' 'Yes,' he agreed.

We climbed into the Bocian, me in the front, and he commanded the sixteen men to run. With great shouts, like a Dads' army film of bayonet practice, they hurled themselves forward and vanished over the crest of the hill. The two bungies grew thinner and thinner until something had to give. Something did, and we shot into the air to the accompaniment of a loud bang. As soon as I reckoned we were clear of the hill face I turned over the trees towards the lakes, with the glider shuddering in a very strange way. We were also sinking down faster than I expected, even though this was the first time I had flown a Bocian. Now I was over the lakes, but kept turning around the mountain as we did not have much height to spare. Not a word came from Adam, and I was too preoccupied to start looking round to see if he was all right – he was, after all, the chief instructor. To my great relief the flower field appeared, and I could see the elegant little winch station with its beehives in deep shadow. Landing was now no problem and once again we rushed through scented blossoms, and stopped. Then I looked round. Adam was sitting in the back seat with both arms in the air, holding the rear canopy. Apparently the acceleration produced by the sixteen men had been so great that it had jumped its runners. As it went Adam had caught it before it took the glider's tail with it. He grinned, got out and carefully fitted the canopy back. Then it was again up the mountain, with the Bocian on the flat car in the last gleams of evening – and supper.

13 · No pesetas in the pampas

1963 was a terrific year right through. It started in early January when Liz, Jan and I went to Davos for two weeks' skiing. The sun and snow were perfect but already the temperature was giving warning of the sort of winter it was going to be; for the last few days it was skiing on almost deserted runs in a temperature of −30°C with freezing smoky breath.

Every World Championships I went to was an adventure in itself, but particularly so when it was a new, and to me, unknown part of the world: even more it was a delight to be in southern hemisphere sun while our home was deep under snow in the worst winter 'since records began'.

Argentina was a delight, but to us northerners it was almost permanent though enterprising chaos; what the organisers lacked in preparedness they made up with willing and very *ad hoc* improvisation.

Before driving the 140 miles from Buenos Aires to Junin, a nineteenth-century railway town far into the pampas; originally built by the British and remarkable for tall sandy-haired Argentines called Miguel and short, swarthy ones named McGregor, we had to get ourselves mobile and extricate our trailers from the shipping line after their winter journey across the Atlantic. In spite of help by Shell and tireless work by John Furlong, it took three days and the British Air Attaché, Group Captain George Lerwell, to get our cars and defeat bureaucracy. But we were lucky; some teams did not manage this difficult feat until almost halfway through the practice week. Then we had to learn to drive in Buenos Aires, which we certainly had not appreciated was the size of London and used largely by locals who knew the way, so signposts did not need to be repaired or even exist. In our innocence we wondered why every car had enormous bumpers, until we stopped at a red traffic light, and were only just avoided by the stream of hooting traffic which poured past. But mostly the days in the heat were fully occupied repairing damage on the trailers and tightening glider control cables which hung in warm loops after their life in cold, damp England and at sea. I also had to create a rubber stamp out of a potato to validate John Willy's international driving licence for trailer towing which some official had missed doing.

The British team arrived at Junin a few at a time due to car renting problems and because Nick Goodhart's Comet had been delayed in Madrid due to snow. We had been allocated for accommodation two small summer weekend chalets in a wood between the airfield, and a vast and beautiful but rapidly shrinking lake. Two chalets were clearly going to be too cramped for thirteen of us so I joined the queue of other team managers with the same problem: solved quite soon by the organisers persuading more townspeople to loan more chalets. Although from the noise of hammering on the airfield it was quite apparent that the organisers were still a long way short of readiness, our welcome was great. On the first official practice day at eight o'clock in the morning four of us from each team were installed in a fleet of twenty-three Citroën 2-cvs, standing like vertical sardines through the sunshine roof orifice, to be driven slowly in procession through the town. We were totally unprepared for a welcome by the 76,000 people of Junin who crowded the streets waving and cheering. The French team immediately in front of us with rhymthic *joie de vivre* bounced their Citroën up and down so hard the bumper hit the ground and the Argentine driver – probably the owner – jammed on his brakes to remonstrate. Our quick-thinking driver did the same, almost propelling us out through the roof in the middle of our bowing 'Queen' act.

After an exhausting three hours we were driven back to the airfield to find the Chileans had arrived with everything in their two two-seater Blaniks, having been towed by Piper Cubs 16,000 ft over the Andes. One Chilean pilot was also called Williamson so we christened him Chillywilly to distinguish him from John Willy. We had also been allocated our third chalet, and electricity had been installed in all of them so we would no longer have to go to bed in the dark.

The practice period was not dull for any of us, particularly the Japanese pilot, Oda, who had his appendix removed in the local hospital. Nick started to learn about local retrieving when he landed 25 km short on a 328-km practice goal race in country apparently empty except for cattle. He hurried to find a gaucho or two to de-rig the Skylark and take it to a farmyard before it was eaten by inquisitive bulls, then he went by tractor on a compass course 3 km to the property boundary, climbed the fence and walked 5 km, also on a compass course, to *the* road, hitch-hiked to the nearest village – the fifteenth vehicle stopped for him – discovered the village had no telephone and hitch-hiked 35 km in the opposite direction. Fortunately he now found both a telephone and excellent hotel, and was retrieved next morning.

Work that we could have well done without was cleaning the glider and ourselves after the airfield had been heavily sprayed with DDT at low level and without any warning. Another job we had was to erect our 48-ft base aerial, so that we could keep contact with pilots for as long as possible. This had been designed and built of wood by Lorne, was very elegant, and had travelled in one of the trailers. We laid it all out on the ground and pulled it up by its slender nylon guy ropes, whereupon it gently bowed from the waist. After a long and highly technical discussion we added more guy ropes with someone at the end of each, installed John Furlong on top of a convenient fire engine some distance away, holding an extra steadying line from the top, and got Gerry Burgess to drive slowly towards the aerial with Lorne on the car roof carefully pushing it up with a rake. No one offered to help although spectators multiplied by the minute – at a safe distance. At last we got it up, except that it now hung its antenna head like a wilting sunflower. After further ideas and more nylon rope we finally got it properly set up and went off for some much-needed beer; but when we came back it was doing its sunflower act again.

We also had a lot of unexpected unserviceability problems. The radio sets had been overhauled wrongly and although transmitting all right, received very little; and Nick and John Willy's leading edges kept bursting out in bubbles. Several times they had to be stripped down and filled and refurbished which was extremely tedious work in the heat and constantly blowing dust. But we were lucky with people. Apart from our own ever-willing team, a wandering New Zealander, Warren Spence, attached himself to us, and worked so hard that we quickly adopted him officially. We were also lucky in that the owner of our chalets decided to adopt us, keeping us supplied with excellent drinking water, and one evening giving us a barbecue party. The family arrived from the town complete with silver knives and forks, glasses, champagne, salads, home-made pies, and, of course, huge chunks of beef to cook on the chalet's permanent charcoal grill. With little common language I hope we got across to them our grateful thanks, as it was a most welcome and civilised interlude.

The first contest day was on 11 February, and resulted in so much confusion and glider damage that the organisers declared the 12th a rest day for reparations! Having by a miracle avoided any breakages, we spent the day sunbathing at the golf club – all competitors being made honorary members – and working out plans for survival.

The problems of the 11th had begun with briefing, which was quite inadequate as to start and finish line at the intended goal at Mercedes, 141 km away. Anticipating this, due to the absence of any supplementary regulations and information, we team managers, including again the forceful Pirat Gehriger, had been trying to forestall crisis, but with no effect. During the hubbub someone discovered that the latitude and longitude of the goal was not where the airfield was on the half-million map, so a large-scale map was produced, but this wasn't correct either. But once out on the field the launching, by a marvellous mixed bag of Stearmans, Fleet Finches and other somewhat traditional aeroplanes, went well; better in fact than the opening day display of glider formation and inverted flying, aerobatics by a Meteor 4, with the simultaneous release of thousands of pigeons for peace – all but one escaped the Meteor – a Cessna 210, and aerotows by helicopter all at much the same time and at no noticeable altitude. There was also a parachutist who landed on the live electricity cables which fed our chalets, riveting attention as he climbed nonchalantly down from the pylon, his canopy still attached and draped over the wires.

All pilots reached the goal at Mercedes airfield with its grass runways, but no one had been told that all the 'grass' between them was 8-ft high maize or sunflowers, or that there would be no landing T or any control at all. As a result, gliders came in from every direction and several were damaged, mostly by running into the maize to avoid each other. But this was nothing to getting them back home again. At Junin all retrieving was by aerotow by very enthusiastic Argentine pilots, so it was not long before the many tugs landed on all three runways at Mercedes, and started towing off the fifty-two gliders. Control was now by a man at the cross-section of the runways who waved his hat to the tug he wanted to go. From the distant pilot's seats, however, any wave was an invitation to go; so they went, John Willy's tug leaping over Tony Deane-Drummond at ducking height. But in spite of the enthusiastic haste several gliders still had to be landed in the dark at Junin by the light of a few flares.

Another problem becoming evident was that the aeroplane tow hooks were not very reliable, so the tow rope was inclined to fall off the tug end, usually during the climb-out, followed by its back-releasing from the glider and being lost. This happened to Nick on the way home, so he and his tug landed in a large field. He had with him the ordinary 20-ft field rope, which each of our gliders carried, so that wandering gauchos could pull the glider out of the way of cattle;

being a sailor he promptly unravelled this, tied the strands end to end to make it longer, and lashed it firmly to the tail of the tug, keeping the quick-release rings at his end. It worked. Two days later John Willy had similar trouble. There was usually some delay in the tug reaching the scattered gliders, as without radio the roving aeroplanes were dependent on the pilot's eyes to spot gliders on the ground, and then to land and tow them home. To help speed this process each pilot was supplied with a pocket heliograph by which he might just flicker the sun into the tug pilot's eye. On this day, John Willy had been waiting some time when a tug appeared and landed, only for the pilot to announce that he was already committed to another glider. After a while a Stearman turned up, the pilot shook his head at the field surface full of bumps and armadillo holes, but agreed to try. They bounced out of the field all right, but at 30 ft over the trees the tow rope fell off the tug. All John could do was turn down wind and land heavily in the next field. However, nothing was broken in spite of the even greater frequency of bumps and armadillo holes, and tall spear-like thistles. (I supplied each pilot with self-adhesive patches for get-you-home use.) The tug pilot tried again but this time the tow rope fell off just after take-off. This discouraged him so much that he just flew away. With the hot sun moving steadily down the sky John was beginning to feel not a little depressed when a Fleet Thrush landed in his previous field. The pilot came over, looked earnestly at the short run and the armadillo holes, sucked his teeth, went back to his aeroplane and flew away. Eventually another tug landed in John's field, the pilot very discouraged by the size and surface – when he had finally stopped bouncing on landing – but he agreed to try. At this John produced his personal rope – since Nick's episode we had cut our spare coil of rope into three so that no pilot need unravel his field rope – and when the tug pilot was searching the field for the most hole-free take-off line, surreptitiously tried to shorten the rope. Immediately the gaucho, holding the wingtip, seized it and tied what was obviously an excellent knot, which John quickly learnt. The tug pilot grimaced at the abbreviated rope but took off. It worked and they got back to Junin as darkness fell, with John in serious need of a large beer. The organisers then agreed to *do something* about the tug hooks.

Many of the problems at Junin were, of course, due to never having run a big Championships before, and not realising, for example, that finish lines should never be near the far end of the landing field. We had, of course, pointed this out but nothing happened until after the end of the Out and Return to Pergamino on the third day, when they watched aghast as fifteen gliders crossed the line together and had to land in the small cramped space left to them from whatever direction they came in from. But it was not only the flying which was testing our ingenuity at every moment. Back at base Lorne was at war with the local banks who were refusing to honour traveller's cheques, and had to make a special journey back to Buenos Aires to get cash. Without team money everyone had borrowed pesetas from everyone else which made accounting difficult. Lorne was also duty shopper, as he was best at drawing diagrams of what we needed, including special bits for the radios.

But yet another problem was the laundry service which had been promised but had not appeared – so new shirts and pants featured increasingly on the duty shopper's list. Then one day it happened. 'Today the laundry will be collected,' they said at briefing. 'Please to put your clothes in the plastic bag with the name of your team. It will come back tomorrow.' Three days later it did come back, and I went to collect it. It was beautifully clean and ironed, but all the world's shirts were in one neat stack, all the world's socks in another, and all the underpants in yet another. Even by the end of the Championships you had only to look around to appreciate that all was not yet restored to normal, particularly the Japanese pilot in some enormous North American shirt.

Evidence of inexperience also affected the task setting as the met man, so essential to the success of the Championships, had never come across gliders before: he was the Most Senior Meteorologist and although his dissertations, when we could understand them, on the weather from Patagonia to Brazil were erudite and fascinating, they did not really

No pesetas in the pampas

help the task-setter to know whether he should go for a 100-km Triangle or Distance along a Line.

However, we were now prepared for anything, however unexpected, and had done as much planning as possible to circumvent whatever God and the organisers chose to perpetrate next. On 18 February the task given was a 300-km Triangle and the weather looked good – to start with – but by the time pilots were reaching the second turnpoint a monumental thunderstorm was throwing $2\frac{1}{2}$ in. of rain on everything below. The ground became invisible from 2000 ft and as each glider landed it sank into remarkably glutinous mud brewed up from the normal dust. John Willy, in some far field with his glider immovable and draped in a plastic mac, walked to the nearest estancia, the owners of which shooed away this bedraggled figure as a hobo. Then they discovered he was a pilot, and could not have been kinder, feeding him on English tea and home-made scones. This was better than another outlanding when he was put up for the night at the local hospital and invited to watch a hernia operation before dinner. Everyone landed out on this day and in the mud and torrential rain movement and communication ceased. Most pilots spent the night damply in their cockpits, while others sheltered with peons in little hovels heavily shuttered against mosquitos and any passing air.

Next morning retrieving began but little organisation was evident in the search for the gliders and many were circled by tugs which neither landed nor reported them; even at Junin landing information was not always received by Control. Rolf Kuntz, from Germany, was one of the few to have a tug land in his field – except that the embedded glider did not move, the tow rope broke, and the aeroplane turned neatly on to its back. Then they sent a helicopter for him, but just before taking off the stranded tug pilot also jumped into the S51, which had not been re-ballasted for his weight; so the helicopter crashed, breaking the pilot's ribs. Then a German car was sent from Junin to get Kuntz's glider to harder ground, but in the rush it took away the main rigging pins of Heinz Huth's Ka6, so a second car was sent in search of the first. Our pilots were luckier, Nick and John reached Junin before dark and Tony Deane-Drummond was put up in a civilised flat; but it was becoming apparent that any day with outlandings was inevitably going to result in the next one being a rest day. This thought was also apparent to the organisers who were continuing to postpone the mandatory Free Distance task in case it resulted in a premature end of the Championships. But with only two days to go, and as good long-distance weather passed by, they were being pushed by the competitors. Apart from the fact that a lot of pilots still liked this long task for its own sake, for many from weak-weather countries Free Distance gave a rare chance to break national records.

So, after a successful 300-km triangle, the organisers took a deep breath and gave Free Distance, briefing firmly that unless gliders kept within certain sectors to the north there was no guarantee that they would be found and retrieved in time to go home as arranged. Because big distances were expected, the tugs would be dispatched to airfields along the likely track so as to retrieve gliders into them; from there they would be towed back to Junin by other tugs.

Take-off was more or less uneventful except for the standing dust cloud stirred up by propellers into which everyone rapidly disappeared – one Stearman was taking off only to suddenly discover another aircraft just in front. The pilot had jammed on his brakes and gone over on his back, but remarkably without breaking the propeller. Willing helpers got it the right way up, and without any check for damage, it started towing again; maybe the Argentinians knew something about aeroplanes that we didn't. For us at base it was a quiet day while everyone fought their way towards the north. I spent some of it on the peaceful shores of the great shallow laguna by the airfield, throwing a boomerang with Australian Jack Igguldon, which was a pleasant change from the noise and dust of the airfield.

By bedtime that night all our pilots were down; Nick at 500 km and John Willy and Tony Deane-Drummond with a large number of others at over 400 km, but already two little-known pilots, Webb from Canada and Baeke from Belgium, were over 600 km away. Results were not known finally until late the following afternoon when Dick John-

No pesetas in the pampas

son, USA, and Edward Makula from Poland in the Zefir, were located over 710 km distant towards Paraguay – they and their gliders came back in a Bristol Freighter; but there was still no news from Dick Shreder, and we were reminded of his landing in East Germany from Butzweiler. Eventually he was found in the ill-advised swamp area 630 km away, having, as usual, glided on until the ground arrived – this time ankle-deep mud in a huge expanse of nothing. Eventually, having been chased by a bull, he found an Indian family living in a tumbledown hut. He spent two days with them, trying to mend their car after he and nine Indians had manually failed to move the glider. In the evening they fed Dick a special treat, roast dog, which gave him dreams that it could have been the family pet. Finally the glider was pulled out by two horses and moved to a dirt road hard enough to tow from – in the usual cloud of dust. Meanwhile, of course, the Great Retrieving Plan was in full swing – except for the navigation. Towing south, the tug pilots seemed to mistrust the compass, preferring to go by the sun. Unfortunately this kept moving. One competitor, Ted Pearson, on his way to a collection airfield, realised his pilot was lost and unsuccessfully tried to steer the tug by flying out to the side to swing its tail out with him; but it didn't really work and in due course the tug pilot landed, and so did Pearson who showed him where they were. Finally reaching the collecting airfield he was hitched on to another tug to take him south to Junin. It was a long tow so the sun moved quite a way. In due course Pearson saw Junin far to the east as they flew sedately past it for another 100 km, when the tug ran out of petrol. This time Pearson didn't land, but found a thermal and soared back to Junin to tell the organisers where they would find their tug pilot.

So, as the excitement of this long task died down, all that was now left was the prizegiving jollifications – with quantities of enormous trophies going to Makula of Poland, and Hüth, flying his favourite Ka6 in the Standard Class – and the inevitable procession in the little Citroëns. This time it was so that we could all place wreaths at the memorial to a national liberator, Bernardo O'Higgins. As usual there had been no advance warning of this and every country was left to solve the problem of an instant wreath overnight. I had no more idea than anyone else so rushed into town to 'our family' who immediately dropped everything they were doing to help. On this final procession we were again four standing up through the roof of the 2cv but also clutching the enormous wreath, which Nick, Tony, John Willy and I held in turns as it was remarkably heavy. Having finally paid our homages we returned to the 2cvs to find that each car now carried a national flag fixed on a tall post. Our Union Jack was of course upside down but too firmly nailed on to do anything about it, and we were driven at high speed back to the airfield for another flying display – with bare inches of clearance between the top of our flagpoles and the drooping electricity cable to our chalets. If the French had done their bouncing act again they would have fried in front of us. This time, as well as the low-flying Meteor, gliders, helicopters and parachutists, there was a concurrent display by gauchos on horses showing us how to catch baby bulls all over the landing area. And so after a final happy barbecue we set off home, arriving at Buenos Aires airport early because there was only one flight out that day and apparently everyone else was booked on it.

We checked in and noticed that they did not weigh our luggage – so I needn't have packed all the heavy tools in my hand baggage. As we sat waiting other teams came in and no one weighed their baggage either, but as more and more arrived the check-in people started to show signs of alarm – it seemed that they had only partly filled Comets before so weight had been no problem. The latecomers' luggage was then weighed with great care and the owners mystified by little sums being done on scraps of paper. We were philosophical. The good Lord had provided for us admirably over the last three weeks – though He hadn't given us any prizes – and we couldn't believe He would let us down now. Finally we all piled into the aircraft, and as the overloaded Comet rotated into the air there was a spontaneous cheer from the passengers – every one a pilot. As we set out over the Atlantic and back to winter – to the same snow that we had left, now a dirty grey – I knew we wouldn't forget Junin in a hurry. In some strange way, although the chaos had been real and often frustrating, it faded into unimportance beside the enthusiastic welcome, the ebullient hospitality, the sheer enjoyment which the locals had got out of what was to them their own Championships, and the fabulous spacious country under its hot sun.

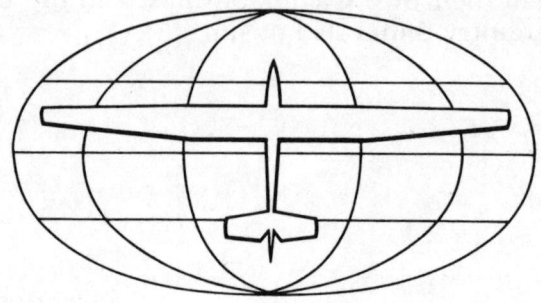

14 · A Worlds of our own

For some time Philip Wills, who was the British delegate on the FAI Gliding Committee, CIVV, had been wanting to make a further bid for the World Championships to be held in Britain. With contests as big and as important as they had now become, the problem was to decide the most suitable site before putting in a firm bid for them in Paris. It was a real problem because any club which took on the job of host would not only lose the best months of summer flying and frustrate its members, but would have to indulge in all manner of works-and-bricks activities, even if they were only temporary and made of canvas. The opposite argument was of course valid – that it gave the host club the opportunity to increase its permanent buildings at either a subsidised cost, or no cost at all. There was also the matter of controlled airspace, and the obvious place, Lasham, had to be ruled out because the London Control Zone was too near and could inhibit competition tasks to the east. After a great deal of discussion and help from the RAF Gliding & Soaring Association, it was decided that the BGA could rent RAF South Cerney, in Gloucestershire, which was in good shape and temporarily under care and maintenance. This seemed to provide all the answers and Philip put in his longed-for bid, which was accepted. I think he had hoped to organise

A Worlds of our own 165

and run the Championships himself and was not a little disappointed when I was given the responsibility. But I was just as determined as Philip that it should be done really well – and by now I had absolutely no illusions about how much effort went into World Championships organisation before the bands played and the flags went up.

Obviously, I was going to need a great deal of help, but not much of this could come from the BGA office, which was already working at near capacity. There was also little money to spare since most of the £9000 sponsorship provided by W.D. & H.O. Wills would be needed for the rent of South Cerney and other site work. I was to get my travel costs for visits there, for stamps and telephone calls, and that was all. Everyone else would work in their spare time for nothing, except reimbursement for the cost of materials for making equipment. It was to be the biggest voluntary effort that British gliding had been called upon to make, and with helpers scattered all over the country delegation had to be made to work. One of the things that I had learnt from other Championships was that communication failure is easy to achieve, and we could not afford such a luxury: so it was essential to work out a master plan, with a clear and simple chain of command: to divide the organisation into Departments, Sections and Subsections, each with a person in charge, who would find his own helpers – and try to avoid inadvertently pinching helpers from someone else. Some Sections, such as Equipment, responsible for setting up a construction shop for all signs and notices and a stores, would have almost all their work before the Championships and reduce to only one or two people during the event, while others would have little in advance but work almost non-stop during the competititon, as, for instance, the group which would produce the Championships daily magazine. All this was written into a Work Book, which detailed what had to be done in each section, with dates for completion: I was determined that competitors should not arrive at South Cerney to the usual sound of last-minute sawing and hammering.

Every Section head received a copy of the Work Book, with spare copies for assistants needing to liaise with other helpers. There was, of course, a Helper Agency Section which recruited volunteers at the 1964 Nationals or through clubs, and put them in touch with suitable Departments; or looked for a specialist helper for a Department which was short-handed. In addition, every helper had a list of names and addresses of all others, plus a check list of who would receive what information. If, for example, the Security and Barricades Section – whose job was not to deal with riots but stop competitors leaving their cars all over the airfield, or the public from walking into propellers – would have their plans for car park locations passed automatically to the sections dealing with glider retrieves, signs and notices, and accident arrangements. This automatic circulation was done from the BGA office using the prepared list. There was also an overall BGA Championship Committee, but it was intended to meet only four times, once at the very beginning to agree policy, and at the end of the three preparatory stages to ensure that all was going according to plan and to consider unforeseen problems – such as the Government's refusal at a late date to issue visas to the East Germans because, at the time, East Germany was not a recognised State.

I think the time spent on this preliminary planning and the production of the Work Book was useful as we ended up with few holes in the organisation as distant helpers could get on with whatever they had volunteered to do without interference or confusion. I think, too, that we were lucky in the time at which we held these Championships. The mid-1960s were years of relative wealth and stability, as well as being before the serious expansion of controlled airspace which would have made task-setting so much more difficult ten years later. Club gliding was flourishing and there was no shortage of enthusiastic helpers, who were also extremely capable.

For me work became full-time and stayed that way, but it was wonderful to find ideas becoming reality and actual hardware. About once a month I visited South Cerney, sometimes on my own to talk to the RAF chief, Group Captain Ottewill, and his team, and sometimes with Department heads. The main on-site objective was the fascinating one of changing a military station into a congenial 'village' where people could find each other without the endless searching common on many airfields; and above all, where friends of any nationality could sit and talk or drink, preferably in whatever sun Wally the Met could arrange. To achieve this we arranged caravans and the big scoreboard to surround a space between the control tower and the fire engine house to make a sheltered village square. The main Championships offices were in the rooms along the side of the hangar just behind the square, and the fire engine house became the souvenir shop. This was run so effectively by Stan and Pat Armstrong of the Derby and Lancs Club that it made a much-needed £600 profit. Into our square, during the Championships, we would place tables with coloured umbrellas and plenty of chairs, so that it would become a natural centre of gravity for the three weeks that we would all be living at South Cerney. But there would also be times when competitors would want to get off the airfield and see something of the local, and beautiful, Cotswold countryside. Remembering how we had appreciated being adopted by 'our' family in Argentina, we asked if any local householders would adopt a National team. Nothing formal was to be arranged, the object was simply to have local friends to visit for a quiet evening off the busy airfield. We hoped that we could find enough families for the countries expected; but we need not have worried. We were overwhelmed with offers and some teams got on so well with their families that they remained in touch for years afterwards.

I moved to South Cerney about two weeks early, leaving our garden with a sort of rain forest appearance, to look after the many final details; such as the setting up of the main hangar 'club' area (for wet weather), and the packaging of regulations, tourist maps, and invitation cards into a plastic wallet for each competitor, crew and key official: 600 in all. We also had to make the final allocation of beds for the teams. There were rooms and dormitories of different sizes on the airfield and the problem was to allot them so that as far as possible teams were together, with pilots having the more restful smaller rooms. I also kept 150 beds spare and not on any list, to cope with the camp followers and helpers who always turn up without notice at any World Championships. It was only just enough; by the official opening day, presided over by Prince Philip in freezing rain, we had not a spare bed in the place.

Unfortunately, the weather was one thing that not even Wally could manage to control and the first week turned out to be one of the coldest early Junes since those famous records began, and as Wally said at one of the three language briefings, 'I'm sorry, but the weather is unintelligible except in English.'

At South Cerney we had almost 300 volunteer helpers. This may seem excessive but some came only for one of the three weeks, and others just at weekends. Glider manufacturer, Fred Slingsby, for example, took on the laundry service with his wife, Fluff. In addition there was enormous support from the RAF, who did all the towing, managing 95 per cent serviceability of the Chipmunk tugs: their simultaneous start each morning was impressive. They also ran the catering (100,000 meals), the small but efficient camp hospital; and the cinema which showed feature and gliding films, including one of early rogallo hang gliders; and did many other things including charging 900 batteries. The RAF also twice repaired the USSR metal gliders after field landing damage. It was not, in fact, only a World Championships but a public-relations exercise for gliding as a whole, and a way of showing Britain to visiting teams. Invitations went out to VIPs, who had to be looked after, and schools were offered conducted tours so as to introduce children to gliding. The *Championships Journal* team produced 1000 duplicated magazines almost daily, with all important items in three languages. Then, of course, there were the thousand and one little happenings such as helping a pilot burgle his own trailer because the keys had gone to London in the team manager's car, or discovering that the Swedes had brought the Norwegian national anthem instead of their own, and had compounded the problem by accidentally driving over a Norwegian wing – fortunately without serious damage – or hurriedly having to make an FAI flag out of RAF bedsheets because the proper one was stolen by souvenir-hunting pilots just before the closing ceremony. Then there was the final banquet itself to which teams, their host 'families' and all helpers were invited – more than 1000 people – with entry by means of the official banquet menu, so as to exclude gatecrashers.

There were only six flying days, not as many as we had hoped, but enough to make it a fair and worthwhile contest. The three weeks were also free of any serious accident, the nearest being a collision in a thermal between Cartry of France, in an Edelweiss, and Petroczy of Hungary in a metal A.15. He received only dents and scratches on one wing and continued with the task, but Cartry lost rather more of his wing. All this happened about 20 km out from South Cerney and Cartry soared home, calling over the radio that he wanted to replace his wing with a spare and be immediately re-launched. Equally immediately I rushed out on to the field in the Land Rover as changing a major component was forbidden unless the damage was not the fault of the pilot, and this we did not know – nor with such a collision might ever know. As Director my difficulty was that if I refused permission until the matter had been investigated, there was no way that Cartry could fly the task and he would lose all points for the day, effectively putting him out of the competition. The Edelweiss landed as both I and the French ground crew converged on it – they even had the spare wing with them. 'Can we change it?' 'Yes,' I said, 'he can fly with the spare wing but "sans prejudice". If the investigation finds he was at fault his score will be deleted.' 'OK, OK,' they grinned, and had Jean-Pierre Cartry back into the air inside ten minutes. As I had suspected, the investigation was not able to allocate blame and Cartry was allowed his points. There was no protest.

With such weather there were naturally many outlandings and a great deal of retrieving. On one day the Station Commander and his Wingco had left the airfield for a couple of hours' peaceful golf far from anything aeronautical, only to have a glider land on their fairway. The Irish team, who had supplied Roy Jenkins with a mug of whiskey when they were introduced to him during the freezing opening ceremony, went to retrieve Tom Evans, who called on the radio that he was in the middle of a lake. Mystified, they arrived at the stretch of water to discover Tom sailing with a local boat-owner, having taken his portable radio with him.

But South Cerney also showed up a serious problem for the future as several teams, including the British, had brought powerful base radio stations. Earlier, radio had been used to assist with retrieves so as to get the pilot back in time for a good night's rest, but inexorably it was beginning to control the pilot. Base aerials now towered 50 ft in the air, and had to be kept away from airfield approaches; direction-finding aerials sprouted to help pilots with navigation; and in a few cases tactical decisions were being taken at base. This was not the idea of Championships at all: the contest was between individual pilots. There was even evidence at South Cerney that one pilot was employing an aeroplane to fly ahead of him to look for thermals. This whole matter would need to be discussed at the next CIVV meeting and decisions taken to control this babel of electronic advice, quite apart from the hazard of doubtfully supported 50-ft aerials!

But this was not the only new or developing innovation. At South Cerney, for the first time in World Championships, we used non-competing gliders to look for early thermals, so as to help us decide the best moment to start launching. It was not only for the competitors that we wanted to get it right. The RAF were launching both classes in two spaced lanes, and getting all eighty-four gliders up in 24 minutes, so for this time the approaches were thick with tugs and tow ropes, and we wanted as few descending gliders among them as possible. Pairs flying – two or more pilots of a team flying together for their mutual benefit – was also practised, mainly by the Poles, and there was great discussion as to whether this was fair, and if not, how it could be stopped; yet another matter for Paris.

The Championships were won by two young pilots: Jan Wroblewski of Poland in the Open Class flying a 15-m Standard Class Foka, while François Henry of France won the Standard Class in his Edelweiss. But in the five years since the Phoenix appeared at Butzweiler only three glass fibre gliders had been built – the Phoebus, the Hütter

A Worlds of our own

Libelle, and the D-36 – and they were all German. All too soon everyone returned to their own homes, and Lorne departed with the girls; Liz, who had been assistant crew to the Russian team, and programme seller Jan, aged ten. I stayed on for a few days feeling flat and to help with the clearing up, deal with the extraordinary range of items from expensive radios to favourite woolly hats that had been left behind, and thank Peter Ottewill who had been so helpful and patient with the mass invasion of his RAF Station. As I wandered round I could still hear all the voices; the crewman with very little English asking Barbara Wrigley in Hospitality where he could pick wild orchids, the RAF officer who arrived breathless to announce that the mobile ladies' loos were now installed and could we please test, Wally the Met saying 'This afternoon there is definite risk of thermals', and Frank Irving, Deputy Director, at briefing on the first hot day: 'The Station Commander insists that no matter how few clothes you are wearing, you must wear your name badge'; or Chris Riddell over the tannoy: 'The following flotsam has been washed up in the Information Office: car keys Union FP638, pair of Fair Isle mittens, gold bracelet, notebook belonging to beer drinker with contacts at Aldershot and Tulse Hill,' and not least the crackling cacophony of the Chipmunks starting up, and overhead the soft and gentle whistle of circling gliders. Then I, too, went home and the lively human world that was our own World Championships slowly faded into the past.

15 · A different sort of competition

After thinking only about South Cerney for almost two years I felt in need of a change. I wanted to spend more time with the family: Jan was coming up to eleven and was working to get into grammar school. At Lasham I wanted to do more flying, and I needed more time for painting and writing books, as well as running the small *BGA Instructor* magazine. I was also coming up to fifty and had been vice-chairman of BGA for seventeen years. Philip was still chairman and likely to remain so for a long time so, to give myself more flexibility, I resigned the vice-chairmanship; this would also give some other Council member the chance to 'move up', but what it actually gave me was yet deeper involvement in World Championships, now through the FAI.

During the preparations for our World Championships I had gone to meetings of the CIVV in Paris to discuss our rules, but was now invited again to discuss the problems of sophisticated radio stations. Philip Wills was the British delegate at this time with Pirat its long-standing President – though it took time for me to get used to glider pilots wearing suits in a conference room instead of flying overalls at a windy launch point. From then on I continued to go to Paris becoming the British delegate, and working on the

A different sort of competition

updating of competition rules so that they kept pace with the sophisticated way in which gliding was developing. When agreement was reached – sometimes a lengthy process – the rules had to be put into words that could be translated into other languages without ambiguity. English was used and both Pirat and Piero Morelli, from Italy, knew this language so well that they were great to work with; I learnt more about the concise use of English in Paris than I ever did at school.

One reason I think I enjoyed word drafting and, for that matter, other writing was that I have always found it easier to put my thoughts and ideas on paper than to explain them verbally. Instructing a student pilot or spontaneous discussions were fine, but I could never manage to say quite what I wanted on formal occasions, and so felt bothered by them. It took me until I was really quite old to more or less overcome this inhibition, though I have never become skilful.

It was not only apparent from working on the rules that gliding was moving into a period of substantial change. Many manufacturers were now building glass fibre gliders which would demand new techniques and skills from the pilot, which in turn would need training methods tailored to suit. It was my responsibility to see that these new pilots were properly taught, and for this we were about to get some unexpected help.

Since their generous support at South Cerney, W.D. & H.O. Wills had continued to assist gliding with instructor scholarships, and now offered to sponsor a national competition organised by the BGA Instructors' Panel to find the best trained club pilot. The winning two clubs – in the north and south – would be given a Slingsby Swallow glider. Pilots who entered had to have a small amount of solo flying and be under forty. Each received a training manual while the Chief Instructor of their club was sent a syllabus, to be covered in eight weeks. At the end of this period they were sent test papers to help them choose their own best pilots, which produced sixty-one semi-finalists from forty-seven clubs. CFIs were then sent a second eight-week syllabus covering various subjects from precision flying to towrope splicing, with the semi-finalists coming to ten centres to be tested by examiners of the Instructors' Panel. The best from each centre would go on to Lasham for a three-day session of tests. There, each would fly with two examiners: National Coach John Everitt and Doug Bridson of the RAFGSA, who pooled their scores; and then be tested by Harold Drew, who had a parachute for inspection and a heap of batteries and switches to connect to an instrument panel. After that Lorne had them for navigation, and Ray Stafford-Allen for glider airworthiness. Finally, there was an interview and a fine lunch while results were being worked out.

It was a tough marathon but the prizes were worthy of it, and the winners were Michael Barker from the Derby and Lancs Club and Noel Ellis from the Cornish. W.D. & H.O. Wills felt, happily, that this competition was sufficiently worth the money and effort to repeat in 1968. This time 155 pilots were entered including seven women, and all was going smoothly until the prize Swallows could not be delivered on time. With barely a week to go this was a real challenge and the only way to overcome it was to try to buy gliders from Germany. High-speed telephoning produced 2 Ka6s, higher performance than the Swallows, but still good for intermediate pilots; so after a quick paperwork session with Customs, Frank Irving and I set off with a large trailer, hoping our calculations were correct that both would go in to it. Due to the ingenuity of Schleicher's workmen they fitted. That year Gillian Howe of the Cotswold Club and L.J. McKelvie of the Ulster and Shorts Club took the gliders home.

In 1969 I became part-time adviser on aviation sponsorship to W.D. & H.O. Wills and we also ran a similar competition for aeroplane clubs with a Cherokee as prize. They also gave me the chance to make a film about gliding; a new departure but an exciting one. It took me no time at all to decide not to make a standard public-interest film. I wanted to portray the delights of soaring from the pilot's view, so produced a script for an out-and-return cross-country flight showing the pilot searching for good clouds, getting low, wondering where he would land, finding another thermal, reaching his turn point and finally making it back home. With the script approved I set about making the film

A different sort of competition

with John Everitt flying the Capstan, as its side-by-side seating allowed photographer David Lomax to shoot from inside the cockpit looking out. One thing I was determined about was that not only should the skies match – not cumulus appearing and disappearing like a conjuror's act, but that the landscape should also match. In early summer fields were green but in July they were rapidly turning yellow, so it was essential to grab at every opportunity to quickly get all the footage needed. Then there were long sessions in London with a most competent young editor technician. For days, it seemed, we stared into the little screen, stopping, cutting, starting, splicing bits in until finally I was satisfied – but only until I saw it on a big screen. Then suddenly into shot came a ghostly aeroplane where none should be. It was the towing aeroplane flying into the sun. On the tiny screen it was invisible; so back we went to remove the ghost and find some suitable cumulus clouds to splice in. The film was called *Glider in the Sky*. It received a 'highly commended' at an Italian film festival, and a few copies are still around.

While all this was happening the BGA was in the throes of changing its constitution, and among the new rules was one to limit the holding of office for more than five years. Philip announced that, in any case, he was retiring as BGA Chairman, and since I had now been Chairman of the Instructors Panel for twenty years, this would mean retirement for me too. Roger Neaves, a very active Panel member, took over, and so that he would be able to do things his own way without an 'old hand' peering over his shoulder, I left the Panel completely. It was sad to say goodbye to what had been my baby, but it would give me just a little more of vanishing time for instructing at Lasham – and preparing for the World Championships at Leszno for the second time. Nick and George Burton would fly in the Open Class, with John Willy and David Innes in the Standard.

The Championships in that flat, green land where we had shared thermals with the storks years before had been so much fun. But 1968 was not. Roman and Irene were no longer there. Soviet troop movements across Poland limited the area over which the 100 foreigners were allowed to fly, pervading the whole event with uneasy political overtones. Cloud flying was prohibited, and the wet and miserable weather gave few opportunities for the many new glass fibre gliders to demonstrate their much-publicised jump in performance – and sometimes their flutter; although they were strong enough the early glass gliders were not always sufficiently stiff. On the sixth day of rain Frank Irving and I decided to introduce a little light relief into the gloom-ridden atmosphere by turning up to Briefing in wet suits with fins, mask and snorkel, but after 50 hours of non-stop rain the airfield and tents submerged into a total lake, and the Danes had to keep a trench-digging party continually in action to prevent their beds floating. On the plus side the Italians had brought a caravan containing 600 bottles of wine, and as Team Manager I had the use of a Silver Shadow Rolls, kindly lent by the factory. One day we had left it in a small, poor village to do the interminable telephoning and when we returned found it being lovingly stroked by an old man. When he saw we were not angry he explained that he had not seen a Rolls since his Squadron Leader had one during the war in England. We took him for a little drive and he was so happy.

Suddenly, on 13 June, the sky cleared and the wet climbed back into the clouds. Launching started, and, having seen our gliders safely across the start line, we were eating our packed lunch – no 'keks' this time – when Nick reported a collision over the radio: he could see only one parachute. I passed this news immediately to the organisers, who were on a different frequency, and who sent off a towplane to investigate. It had been between Elke of East Germany and Aydogan of Turkey, back in World Championships for the first time since the rudder incident at St Yan. Both had parachuted, Elke very low as he had to delay opening until the bits of his glider had drifted clear of him. He landed in a bog losing his shoes. Aydogan baled out at 4500 ft in a thermal and was still airborne in the upcurrent when the towplane arrived having flown 18 miles from Leszno! Both pilots were forbidden to fly again, and the East German team manager was prevented from speaking about the collision at the next briefing, which we all thought

A different sort of competition

was unnecessarily high-handed.

Then there was another collision, this time between Schubert of Brazil and Ehrat, Switzerland, in the middle of a gaggle of twenty gliders. Ehrat was rammed from behind but continued on for another 110 km with a seriously damaged tail, while Schubert, with no nose left to his glider, landed. This was a repeat of a recent collision in our own Nationals, with one pilot continuing to compete in a structurally weakened aircraft, while the other landed as soon as possible: obviously even more food for the next rules meeting in Paris. For some reason the pilots involved in the second collision were not disqualified – maybe the organisers thought they might soon start running out of competitors. The two crews spent all night mending the gliders so that their pilots could fly next day: Schubert did, landing in a bog; so his crew had a second night without sleep trying to get him out of it.

The weather had by now become extremely hot with great sultry thunderheads all around, but with cloud flying prohibited the storms could not be used. Barographs had to be carried to enforce the rule. On 19 June the storms brewed even before the first take-off, so the task was cancelled to the deafening sound of hail on the roof and the sight of happy Americans running around picking up ice for their drinks. The next day everyone was sent off into the storms, John Willy hitting an invisible telegraph wire while landing close to the gust front, and being in severe risk of decapitation, but he ducked as the wire went through his canopy to cut deep into the headrest. His Slingsby Dart arrived back at the airfield workshops to join what had now become the social centre of the Championships, there were so many gliders being repaired. Our Mike Fripp was the star of the night with a crowd round him for the pleasure of watching such a highly skilled craftsman in action.

Meanwhile we team managers were trying to sort out the inevitable problems caused by the barographs and to decide if anyone had entered cloud against the rules. Each time a barograph was calibrated a different answer was obtained, and in any case no one – not even the Met man – knew the exact height of cloudbase at the time. So after several hours of hassle all pilots concerned were given their points.

The Championships were deservedly won by Harro Wödl, an Austrian partly crippled by polio, with American A.J. Smith becoming Standard Class champion. Our best was George Burton at seventh in the Open, but this time our poor placings were no fault of the pilots. Since Fred Slingsby's retirement some years earlier the firm had changed. None of the gliders for the Championships had been delivered on time: Nick's HP14, built under licence, was not up to the expected performance; and when the trailers were finally delivered a bare week before departure they were still unfinished. As a result, practice flying had been seriously affected by the uncertainty and lack of equipment: at the last minute George Burton had given up the struggle and borrowed a German-built SHK. It was becoming sadly obvious that no longer could we fly British if we wanted to have any chance of winning.

On a wet and windy 26 June we returned home across the Channel to Dover to discover that there was a rail go-slow; so we stopped near some sheds to swap passengers and luggage so as to get everyone as near to their own homes as possible. In the middle of a remarkable scene of confusion with suitcases, radio transmitters, cigarettes, bottles, cameras and binoculars strewn all over the road for restowing in different cars, the police arrived. They told us that the quiet place we had chosen was under television surveillance by Police and Customs and would we kindly not be too long. We knew then we were home.

Much of that summer Lorne had spent designing a boat. This was nothing new; he had been designing his dream boat for years, having built and sailed a dinghy pre-war and won a yacht design competition when a POW. But after Leszno the idea of sailing on the sea far from committee meetings and thunderstorms was attractive. Over the previous 20–30 years we had built canoes and dinghies for the children and played with them on a local lake, but there had never been time to think about serious sailing. There was no more time now, as we were still both fully committed to gliding, but sailing had a sudden freshness. The only thing that bothered me was that I did not see Lorne's design coming to fruition simply due to this same shortage of time. If he, or I, were going to do any sailing, we had to buy a boat. But it was not only time that was short, money was as well – so to start with it would have to be a very little boat. But there was nothing the matter with that provided it sailed well and fast. One boat around in the late sixties was the Westerly Nimrod, with a centreboard, and a trailer so the complications of mooring were minimal: and it looked right like a good aircraft looks right. The cabin was even more minimal than the mooring problem but it would do for an occasional night. I decided to look out for a second-hand one and when it turned up – sail No. 109 (like the Messerschmitt) – I bought it for £600. We called the little Nimrod, previously nameless, *Sula* – the generic name for that superb soaring sea bird, the gannet – and during that autumn and on every spare day the following summer we sailed little *Sula* with great delight. The challenge, the problems, and the pleasures were so like those of flying; it was just the time-scale that was different. Most things, but certainly not all, happened at a rate a little more in tune with our now advancing years.

Another new venture for me, and a delightful one, was to start diving. This was all Frank Irving's fault, when a group of us, including Liz and Jan, one August shared a flat in Malta; and it led, the following year, to a further holiday at Kyrenia, Cyprus, with Lorne and Frank, when the diving was a dream – at 85 ft it was warm enough in a swimsuit, and clear enough to take photographs. Maybe it was because time was now slipping by too fast that new and different things were becoming increasingly attractive, particularly if they were beautiful to look at.

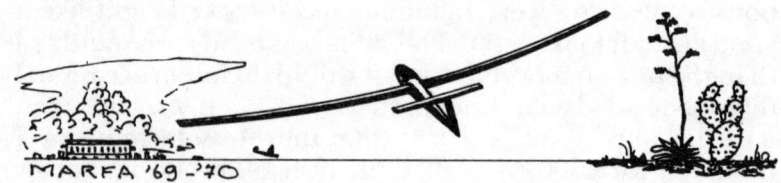

16 · An empty corner of Texas

Subtle changes were now starting to bite deep into gliding's twenty years of stable growth, and the desire for them seemed to be increasing, though not always from need. It was like a snowball rolling downhill. In many ways it was a growing-up, an exchange of the carefree days of childhood for the more serious concern of adult life; though the jump seemed to be a large one, almost straight into a solemn middle age. It was not helped either by gathering inflation or the weather – our Nationals anticyclone had now, it seemed, deserted us for ever. In the 1969 Championships we had only four contest tasks in nine days of rain and storms, and it was to be the last of our big Lasham Championships. George Burton won, with Nick only 74 points behind out of a possible 4000, and at the end the team for Marfa, USA, was to be chosen, now by a new voting system among the top pilots. The feeling of dissatisfaction after Leszno had continued, encouraged understandably by newer pilots wanting to get into the team, and this showed in the voting. Nick was dropped, in spite of the fact that he had also recently won the American Nationals. George was chosen, plus three RAFGSA pilots – John Delafield, Con Greaves, and Andy Gough – new to world flying and with German gliders.

An empty corner of Texas

Marfa, Texas, at latitude 35°, would produce fierce thermals from its hot deserts, and, as retrieving in this great emptiness could produce problems for the pilots, it seemed sensible to go and have a look. The American 1969 Nationals provided the best opportunity to collect useful information, and the RAF provided the reality by flying me to New York on a VC10. This was refreshingly different with rearward-facing seats, and instead of the usual plastic airline meals, a Corporal passed round a container of juicy apples – and returned 15 minutes later to collect the cores. In New York I became extra crew to Steve Dupont to travel the several thousand miles to south-west Texas in his large elderly Ford. As we moved into the southern heat it became apparent that the thermostat could not cope with both the trailer and the air conditioning – so characteristically, Steve threw it away. This helped to some extent, but when it became really warm the only way to keep going with the trailer was to disperse the excessive heat into the car – via the heater. However, we all survived the ride to Marfa, which I found a delightful little town of 2000 people in the middle of a vast desert plateau of cactus and yucca. This was ranged around by ancient volcanic mountains ever changing as the day moved on from a clear pink dawn, to an afternoon of great cumulus castles whose purple shadows lay upon the hills. Then in the evenings, as the clouds dissolved, the mountains took on a soft and gentle roundness in the lowering sun.

The airfield had been built in the Second World War with a single enormous timber hangar, and lay deserted until each summer the gliders came: a migration of white birds.

The competition itself was brilliant; the tasks were ambitious and the organisation full of ideas which worked – like the system for turn-point photography. Everyone, pilots and organisers, was on vacation and the holiday atmosphere was fulfilled, for me at any rate, by almost unlimited sun. Gliders landed on empty public roads, and retrieves could be exciting, trying to de-rig before the next truck came along. From the air the land was fascinating: oil seeped out of the ground to lie in black, untouched pools, huge dry lakes glittered with salt, and through this emptiness straggled the tenuous track of the Southern Pacific railroad; just occasionally a mile-long train chuffed by on its lonely journey from Florida to San Francisco. Marfa was a station on this railroad, 25 miles from the next one, because when it was built the engineer and his team used horses, so the stations were a day's horse-ride apart. Since the engineer had no committee to consult, he named the halts himself — Alpine, where the railroad wandered through the mountains, and Marfa, who was a girl in the Russian novel he was reading at the time.

One afternoon John Ryan discovered that I still had not got my Gold height, although it was now ten years since my 300-km Gold distance flight to Angoulême, and, unbelievably, thirty years since my Silver duration at Dunstable. He quite rightly considered this disgraceful and decided to overcome the apparent idleness on my part by lending me his Phoebus, saying that cloudbase was only just high enough and I would need to climb to over 16,000 ft as Marfa was already 5000 ft above sea-level. John and his crew packed me in to the glider – which I had never flown before – with barograph and plenty of oxygen, and off I went with instructions to release no higher than 600 ft. It was not difficult to find a thermal near the airfield but I could not get the undercarriage fully up, so decided to leave it locked down: the last thing I wanted to do was to be unable to lower it for landing and rub off the bottom of John's shiny Phoebus on the hard ground. The extra drag would not matter too much as, to climb, I would not be flying fast.

As I circled in strong lift with the audio variometer singing, this incredibly beautiful country spread wider, dappled by dark cumulus shadows slow moving over its tawny surface. At 15,800 ft I reached the base of my first cumulus and the lift died: but not far away was another cloud still living. It gave me 16,000 ft before it too collapsed in vaporous rags; and a third one gave me no more. I had not yet lost hope but it was late in the afternoon and the sky had begun to sweep itself clean in readiness for another sparkling night. While thermalling, the Phoebus had also been slowly drifting so the airfield was now 6 or 7 miles upwind, easily recognisable with its broad concrete runway and huge hangar. Below were the hills of Alpine with the sun still hot on their slopes. I was just thinking that this was a good place to be when the air danced a little, and I quickly turned into a lovely surge of lift, looking up to see the first milky vapour of a new-born cloud: even as I watched it grew into little cloudlets soon merging into a growing cumulus. The lift came stronger and I circled carefully – probably with unnecessary care – as I knew my only chance was with this bright cloud. Steadily consuming John's oxygen I watched the altimeter needle move past 16,000 ft, then more slowly past 16,200 ft, but with the Phoebus still climbing. As it reached the magic 16,300 ft of John's calculations, the needle slowed gently to a stop, but I continued circling determined to make sure of every foot there was. It became suddenly cool as the shadow of my cloud spread overhead, and with a soft sigh from the audio, the variometer needle dropped back. That was it, only the barograph knew now if my Gold was at last complete. I straightened the Phoebus towards the airfield, flying slowly down into the warm desert air, looking out at my long and silky white wings, and feeling happy. As soon as I had stopped rolling along the concrete, John arrived, chided me for not persisting with the undercarriage, and swept the barograph off to the control office in the hangar, while I parked his precious Phoebus. He came back grinning. 'Well it wasn't by much,' he said, 'but you made it.'

As soon as I got home to England I prepared a report for the team only to discover that during my absence they had chosen a new manager; no more British gliders, then out with Nick; now me. But I really could not complain. Twenty years as Team Manager was a good enough share for anyone. I knew things were changing, and over the years I had pushed for change as much as anyone, but it would be sad not to return to Marfa. In the event, I did. At the Autumn CIVV meeting in Paris the American delegate Bill Ivans proposed me as Chairman of the International Jury, and I was elected. So this was a new challenge to look forward to, though Marfa 1970 was run so well operationally that we had no protests and I had no work to do. But this was perhaps as well as I had also been asked to be Editor of the

World Championships Journal and this barely left a spare moment.

Lloyd Licher, chief executive of the Soaring Society of America, and his wife Rose Marie set up our office in one of the rooms along the side of the hangar, and we sawed the door in half, so we could keep everybody out, but hear the news through the open upper half; though none of this disturbed a pair of swallows rearing their young in a nest just over the door. The intention was to print 1500 copies each day, mailing 500 to subscribers, so the little office was pretty well filled with paper as well as us, but it all worked, with a steady supply of contributions from competitors and young American helpers, who all seemed to carry portable typewriters in their cars. But no editor should sit only in an office, and I longed to do some more flying – any flying – over this land with which I had fallen in love. My chance came with Marfa Air, a Cessna Skylane used as a radio link and search aircraft. Because of the vastness of south-west Texas there was a real risk that some gliders would land where there was neither habitation nor roads. If the pilot was then unable to make contact by radio, which could easily happen with VHF among the Carlsbad mountains to the north, he could be there for days. It was the first contest day which highlighted the need for a radio link/search aircraft, when the German pilot, Walter Norbert, landed his 22-m span Kestrel near the only house he could see anywhere only to find it abandoned. Hungry, he started walking in a suitable direction for roads and people but when, after 20 km, he had found neither he trudged back to his glider and tucked himself into the cockpit for the night. At first light he set off again, to be spotted by one of seven aeroplanes sent out from Marfa. These were private aircraft belonging either to local ranchers or competition helpers. Harro Wödl, defending World Champion, also landed far from people but with his position known to his retrieve crew. Their problem was that the ground was so rocky that they burst three tyres trying to reach him, so he also slept in his cockpit with everything closed against rattlesnakes and tarantulas – although the only person actually bitten by a spider at Marfa was Piero Morelli, in his bed! Late the

23 British Team pilots for the World Championships in Spain, 1952. l–r: Frank Foster, Geoffrey Stephenson, Philip Wills (who won) and Lorne

24 Glider retrieving through a small French town, St Yan, 1956

25 Final adjustments before departing for Leszno, Poland, in 1958: in Horse Guards Parade, London, after tea with Prince Philip

26 Nick Goodhart coming in to land at Leszno in the Skylark 3

29 Retrieve winch flat car at Zar in the Tatra, 1000 ft up from the landing field

30 Irene Zabiello and Adam Dziurzynski, Zar, 1962, during our visit with the Gawron

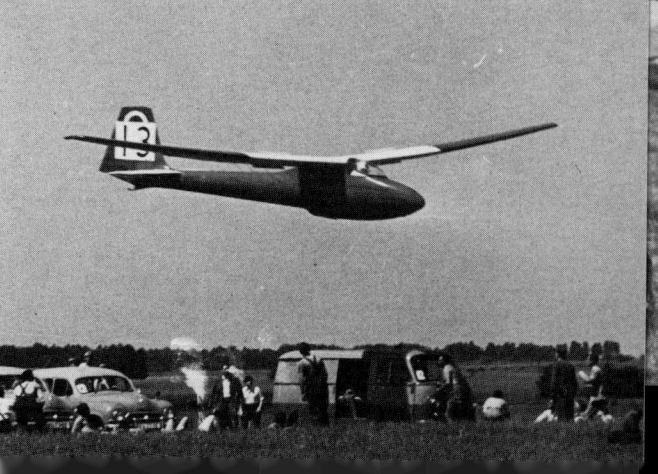

31 British Team at Junin, Argentina, 1963. *l–r front row:* John Williamson, Nick Goodhart, Tony Deane-Drummond, AW, and daughter of 'our family'; *behind:* David Cretney, Gerry Burgess and Wally the Met

32 First take-off at Junin; start of the dust fog

...nce Philip, with BGA Chairman

...k 4 at Lasham: Lorne and Philip

33 Hours of waving from the Citroën 2-cv in Junin town with Tony Deane-Drummond and John Willy

34 Our World Championships at South Cerney, 1965: Group Captain Ottewill behind AW, Prince Philip, Philip Wills and Pirat Gehriger

35 South Cerney: preparing for take-off

36 Our 'village square' at South Cerney. The big score board provided a wind break. Walter (Wally) Kahn in dark glasses

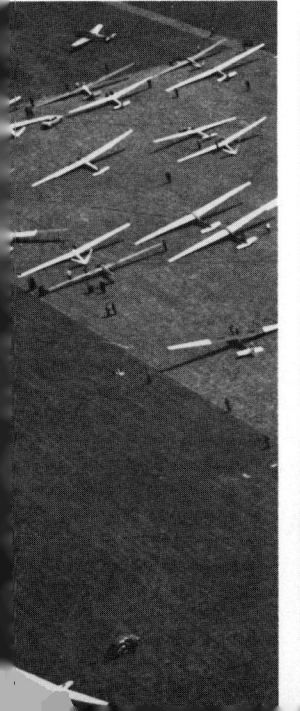

37 The T49 Capstan two-seater during trials at Lasham

38 Testing the Skylark 2 at Lasham: AW waiting to go

39 Vrsac, 1972: the Australians, Wally the Met, Ingo Renner and C. J. Ridley

40 George Lee, Great Britain, the only pilot ever to win three consecutive World Championships

41 Gliding becomes complicated: weighing a competitor before take-off to check that the glider is not overweight with too much water ballast

42 Grenoble, France, 1979: World Hang Gliding Championships

43 Japan, 1981: AW as Chairman of the Jury at the World Hang Gliding Championships

An empty corner of Texas

following day a more agriculturally equipped expedition finally managed to extricate the glider.

After these excursions one or more aeroplanes, usually Cessna Skylanes, were sent out on every afternoon with expected glider outlandings. I went along on many of these trips, mostly flying at 5000 ft but going down to inspect any grounded pilot if we could not raise him on the radio. One evening we saw a glider low, working its way between the bare steep rocks of the Carlsbad mountains, its only possible landing place a narrow road that wound its way along the deep valley. We followed, sheepdogging, as the pilot flew on, and we knew he would not give up until the ground met his wheel. It took him nearly an hour to work through the mountains, only to have to land at the turn point at Ardoin; just as another glider, very low, rounded the point and started for home – even 5 km along the way back would earn valuable points. He just flew straight in the smooth evening air until he disappeared in a cloud of dust on touch-down. We made a low pass and he waved cheerfully, standing up to his knees in the loose dirt. We reported his position as we climbed back up into the sunlight, to be told of yet another glider thought to be near where Norbert had landed on the first day. So off we went as darkness rose around us to search the Sierra Diablo mountains, not for a glider but merely for a light – a flash from a torch, even a match flare. If we saw anything it must be a pilot – there was no one else there. But nothing flickered out of the velvet darkness, so after checking with Marfa base – who had now located the pilot – we turned for home.

That day Marfa Air 2 had been covering the task area to the north and east of the airfield, and on their way home heard frustrated chatter on the radio that George Burton was down inside an irrigation area with his trailer unable to find a way through the ditches to him. In the last of the light this other search aircraft located the gleam of the white glider and then the trailer, and using its landing light, explored the maze of ditches to find a track: then it led the trailer along it in the dark to the glider. At 2230 Marfa Air 2 turned for home where we arrived together half an hour later, aiming to land at a point on the ground where the headlights of two cars converged. Marfa Air 2 was ahead of us when it suddenly turned across our path so as to have a second go at landing, not realising how close behind we were. As probably the only two light aircraft airborne at the time in south-west Texas, it would have been real Murphy's Law if we had collided.

The Open Class at Marfa was won by George Moffat on his home ground flying the great floppy Nimbus II, while Helmut Reichmann, a German schoolteacher in his first Worlds became Standard Class champion. The big changes in the British entry brought no wins, though George Burton did well, coming fourth in the Open Class. Marfa was a championship I was sad to leave. Its remoteness gave it a life of its own, full of enterprise, and it was fun to work with so many good friends. I was happy every moment.

FALKE at LASHAM

17 · Change is nothing new

Ever since the 1920s there had been enthusiasts who put little engines into the gliders of the time. But they had never become popular, mainly because the performance of the glider was not good enough to cope with the extra weight and drag of the engine. Nor was there a good lightweight engine, so the early motor gliders were inadequate half-breeds, not able to soar properly as gliders and too slow as aeroplanes. But by the late sixties such technical problems could be easily overcome, and there was good reason to do so, particularly for training. With plenty of members queues for launches were too long, so that learning to fly was a protracted process: but the motor glider could take off as soon as the instructor and student were ready, and quickly clear the crowded circuit. The only two-seater available at this time was the German motor Falke, but its noisy, low-powered and often difficult-to-start 2-stroke left much to be desired. Sensibly, Scheibe designers realised this and as soon as a Volkswagen engine conversion appeared created the SF-25 Falke around it. There was, of course, opposition by some to anything that would pollute the purity of motor-less flight, so the prospect of powered gliders raised a great deal of steam, mostly from opponents, who gave the impression that at any moment gliding clubs would be submerged under a swarm of buzzing locusts. But there was also a problem with the CAA who wanted to regard the motor glider as an aeroplane and therefore subject it to all the legislation concerned; so it seemed sensible to bring a sample into this country for demonstration, in order that its simple merits could be seen. The BGA managed to get some government money to enable my Motor Glider Committee to acquire an ASK-14, a neat little aircraft with a 25-hp Hirth 2-stroke motor and retractable undercarriage, but with only a single seat.

Frank Irving and I, increasingly practised in fetching aircraft from Germany, went to collect it, and as soon as we got back to Lasham No. 1 Test Group started to put it through a full test schedule, in return for the purchase money. It was an interesting little aircraft to fly, the handling faintly reminiscent of a lightweight Spitfire; the manual retractable undercarriage was easy and simple, but the 25-hp engine produced almost too much power. In level flight it was easy to exceed the maximum permitted speed, so that flying cross country was an interesting process. To avoid going too fast you let the K-14 gain height, if necessary with airbrakes out and undercarriage down. If this was not enough to avoid entering cloud the engine had to be stopped — 2-strokes of that era did not like slow-running and oiled up their plugs.

After gliding down for a while the engine could be restarted by pulling the recoil string — and if this was successful the whole process could be repeated. It was fun sailing across the countryside in a series of gigantic phugoids, and then finally stopping the engine some miles short of the destination airfield to arrive as a glider. As soon as the prop stopped it could be pulled gently round by the starter string so it lay horizontal and least vulnerable to either undercarriage failure or rough gliding club ground. The only problem with the K-14 was the beautifully built Hirth engine with its four mini-carburettors: it was simply not reliable enough, although I had only one engine failure, on take-off.

As soon as we completed the tests and compiled some pilot's notes we let any interested BGA pilot with suitable gliding experience fly it, and by late summer fifty-two pilots had done so: but in August the propeller was broken by an experienced instructor who messed up a simple landing.

As soon as the side-by-side two-seat Falke was in production Lasham bought one, G-AXFR, and CFI Derek Piggott, still running, set about learning how best to integrate it with basic glider training, while I concentrated on negotiating with the CAA to ameliorate the proposed legislation. The CIVV motor glider subcommittee had already approved the BGA's proposals for a motor glider definition, so I had hopes that we could have them treated as a special case – a sort of half glider – as in several other important countries – and could look after them within the BGA. But after three years of meetings the best I could get was a simplified PPL for the pilot, with the BGA in charge of instructors and training. It was better than nothing, but more complicated than necessary and added nothing to safety.

The Lasham Falke quickly proved its value, and a second, G–AYBG, was soon added. The learners' queue disappeared within months, and the club was able to eliminate its waiting list. But argument continued: because training on the Falke was so convenient the student would be kept on it too long, and become a pilot with an aeroplane mentality, and not possess the special approach to flying of the glider pilot. There was, of course, some truth in all the objections, but none which could not be avoided by good instruction: it was essential for the student to learn to judge his gliding distance from the airfield so that he would not get too far away when there were no thermals, and to be subject to the disciplines imposed by a motorless aircraft. I found the Falke an admirable tool for basic training: effect of controls, turns, and the approach to the stall could have time devoted to them that was not possible in a short gliding flight, while simulated cable breaks and circuit planning could be practised more effectively. It was possible to allow the student to develop his mistakes to the point at which he could see the error for himself; in a glider the instructor often had to take over before the fault became fully apparent to the student, so he did not really understand the problem he was supposed to avoid.

Another valuable use for the motor glider came much later, when a pilot was reaching the stage of making his first soaring cross-country flight and needed to learn both to navigate and land safely in fields. With a glider a single field landing could easily occupy a club two-seater for an entire afternoon, including the retrieve, but with the Falke four or five practice approaches could be made into genuine fields in a 40-minute flight. I would set the throttle to simulate a typical glider performance and see how well the student carried out the approach discussed before take-off. If he got it wrong I could open the throttle and climb away, so he could try again. Navigation was even more fun. We would set out on, say, a 100-km triangle for which the student had prepared, and he then had to map read his way around it – not difficult if using the motor glider merely as an aeroplane. Instead I used the throttle to simulate lift and sink. Full throttle was a thermal and the student had to circle just as in a glider. When I closed the throttle he set off on course once more and got on with his navigation. If I left the throttle closed so that in a glider he would have to find a landing field, this is what he had to do. When he had selected a good field within easy reach I would open the throttle again for another thermal, and as the student circled up he had to try to find out where he was. Back at Lasham we would go through the flight and any problems over tea. This sort of flying was for me one of the treats of instructing, and for the student I hoped the start of the delights of cross-country soaring.

It was motor gliders again at the next CIVV meeting in Paris. Could a pilot claim a gliding record on a motor glider provided the engine was not used? Or would it only be a comparable performance if, once stopped after 'the launch', the engine could *not* be restarted? We also had demands for rules for motor glider competitions; how often or how much could the engine be used? So Pirat, Seff Kunz, Chairman of the CIVV motor glider committee, and I agreed to meet at Burg Feuerstein in the south of Germany in June to see what progress was being made in the motor glider rally there, and to draft some simple rules. It was a delightful break in beautiful countryside, living in a castle which had been turned into a sort of hostel a couple of kilometres from the hilltop airfield. But apart from discussions, I was able to get some fascinating flying in various new motor gliders. Burg Feuerstein had only one undulating hilltop runway which at the south end fell steeply into a valley of coloured strip fields, so competition take-offs were effected by all the aircraft taxiing up the runway in line, then turning round so that the last to arrive was the first to go. One afternoon a German private owner with a Falke invited me to fly with him so I could take some photographs. It was good to be in a cherished and polished motor glider instead of the somewhat battered version that school aircraft soon become, but after about half an hour's flying my instructor's eyes, doing their normal monitoring, noticed that the oil pressure needle was in the act of dropping dead. The owner looked disbelievingly at the dial when I drew attention to it, and said firmly that it couldn't be true as he had checked the oil only that morning; then turned towards home, which we would obviously never reach, as an enormous forest intervened. But just behind us was another small airfield, and having no wish to be deposited among pine trees, however fragrantly scented, I again pointed out this nearby salvation. He turned back, entered the circuit and made an excellent deadstick landing on the only runway. Then, so as not to obstruct it for any other aircraft which might want to land, he turned off on to the grass while we were still moving at about running speed. Sadly the grass contained a deep, weed-filled and totally unmarked ditch running parallel and close to the runway. I uselessly shouted a warning as we nosed straight into it amid a great shower of propeller fragments, a flying around of cameras, quite a lot of noise and then silence. We both climbed out but the Falke was too deeply in to move it so we waited for helpers – who were already converging from all directions. They also explained the reason for the ditch – one side of the airfield belonged to the flying club, they said, and the other to military helicopters, and a ditch was less obstruction than a fence. 'Thanks for telling us.'

After some telephoning another Falke arrived from Burg Feuerstein with a new propeller tied to its fuselage. I went back in this one to join Pirat, Seff, and Gerd Stolle who ran the airfield, for another meeting. An hour later 'my' Falke came in out of the evening sky, with its 'dichtung' (gasket) replaced, and wearing the new prop.

Motor gliders had come to stay at last, and I was glad as they broadened the possibilities for uncomplicated and enjoyable flying. They could be flown as a glider to explore an afternoon of thermals, or as an aeroplane to go and visit friends: the pilot could decide.

Whether I liked it or not, and I didn't like it very much, paperwork was becoming an overwhelming fact of life. It was not an accumulation for its own sake but to try, for example, to keep competition gliding free of too many of the complicated regulations which in practice do not always work. The new high-performance glass gliders had brought with them the need for rule changes so as to exploit that performance, but also remarkably fierce opinions from all sides on what the rules should be. The argument as to whether or not flaps should be allowed in the Standard Class continued internationally for three years, and was only resolved finally by having *two* new Classes – one with flaps and one without. The end result of all these changes was that the gliding section of the FAI sporting code now needed redrafting; and I ended up doing it. But it would be wrong to say that I did not enjoy this working with words, even though it was a pity that so many rules seemed to be now needed. Gliding was moving so fast towards extreme sophistication that the simple pleasures of floating in gentle ridge lift in a light glider on a summer afternoon seemed to have been forgotten.

Except in the United States. In the warmth of California a few individuals were building primitive bamboo and polythene craft, mostly with the Dart-like shape of the sailwing invented by Francis Rogallo as a possible means of returning space vehicles to earth, and jumping off hills and dunes. They called themselves the Low and Slow movement; a peaceful revolution against the cost and complexity of conventional aviation. Richard Miller, who had come to South Cerney, was one of its leaders. What he did not know when he wrote his little book *Without Visible Means of Support* in 1967 was that thousands of people were dreaming the same dream – and hoping that it would come true.

I got news of sailwings and rogallos from Lloyd Licher, and sure that it would not be long before some of these basic gliders turned up in Britain I wrote a few articles about what was going on in California for *Sailplane and Gliding* magazine. The enthusiasm of these people to get airborne, if only for seconds, was boundless, and I believed that cheap hang gliding could give its elder relation gliding a needed boost towards more youthful membership. (The average age in clubs was around forty because of the cost.) At first I was hopeful that gliding would see hang gliding as a part of itself, and use its great experience to help it develop. In his retiring speech, BGA Chairman Ken Wilkinson said that hang gliding had great potential in Britain. His successor set up a small subcommittee under my chairmanship to look into the possibilities; but it was probably fear that a new nomadic element brought into gliding would despoil its good reputation that caused a closing of ranks within the BGA, and increasingly a cry that hang gliding was something different; if and when it arrived in Britain it should look after itself. Although this attitude was understandable it was also sad, as on their own these new fliers would have a hard road before them. In *Sailplane and Gliding* for March 1972 I wrote, 'If the hang glider turns up here we should not just pretend that we haven't noticed. On most hill sites there is room for both ... If their owners are not already glider pilots but people starting flying this way, with little more than their own enthusiasm, they will need help, and we can give it better than anyone else. If we ignore them and they have unnecessary accidents any bad publicity is bound to reflect on our type of gliding. Perhaps the Low and Slow movement will never catch on here ... But if it does it will be because these people want something that we once wanted and have now got – just to get airborne and fly.'

But it was no good. Opposition forced my little subcommittee to recommend that hang gliding should grow separately under its own organisation, as the only way it would have a chance to survive. So strong had the closing of the ranks become that there was also less room left for me: why did I want to introduce a disruptive element into gliding?

There was, though, little time to spare for worrying as the 1972 World Championships were about to start in Yugoslavia. Nick was properly back in the Team, flying in the Open Class with George Burton, both on German-designed, British-built Kestrels, while two newcomers, John Cardiff and Bernard Fitchett, were entered for the Standard Class. The airfield at Vrsac was near Belgrade, almost on the Romanian frontier, and the July weather was expected as usual to be hot and sunny with excellent thermals. Except that this year it was not: there were more July thunderstorms since those records began.

But as well as for weather reasons it was a Championships which misfired, as the organisers had not appreciated just how sophisticated gliding had become, and they did not have enough experienced people to cope. This could have been overcome by pulling out talent from among the visiting teams, which would have been willingly given because all that everyone wanted was a good competition, but any organisation limited in experience often seems to extend this limitation to a desire to do everything itself. But when we first arrived on a hot and sunny day none of this was apparent. A large new 'hotel' had been built for pilots and officials which would later become study bedrooms for an aviation school. My room was fine until I took a bath to clean up after the journey. On pulling out the bath plug there was an odd gurgling sound followed by the bathwater jetting straight up out of the loo and flooding the floor. Then I knew we must be prepared for surprises. It did not take long. The opening ceremony air display included dive-bombing and ground-straffing demonstrations by the Air Force. This was most impressive as they used live rockets and ammunition, in some cases firing while diving straight in our direction as we sat on the VIP benches. I wondered if going flat on the ground would be taken as a dishonourable national weakness. One past world champion, A.J. Smith, by no stretch of the imagination either weak or dishonourable, had the courage of his convictions and did, using the bulk of local dignitaries as a screen.

Since my job, as International Jury Chairman, was to deal with protests made by competitors I decided that the best way to reduce the onset of these would be to have the take-off areas inspected for tyre-slashing shrapnel. This was considered unnecessarily fussy but I insisted, so a reluctant squad of soldiers was ordered to comb the two grass strips. They collected over 200 bits of jagged metal.

Another problem, which was no fault of the organisers, but which did not help, was the thieving. Cars and trailers left for only a few minutes would be stripped of mirrors and national flags, and several cars, including two British Range Rovers, were broken into at night and binoculars and radios stolen. The police got most of this stuff back, but later in the competition pilots, having landed in fields, had the cockpit instamatic cameras with their turn-point evidence photos stolen and so lost all their points. Then, on the first contest day twenty-one pilots were given zero score because of bad turn-point pictures. This was a real shock as many of those penalised had an excellent reputation for always turning in good photographs. There was, not unexpectedly, a mass protest; and I spent hours with the organisers trying to find the reason, and discovered that the photo interpretation expert had previously only dealt with vertical shots, and did not understand how to assess obliques. In the end, only three pilots lost points.

The inexperience of the organisers manifested itself not only in the tasks that were set, but in failing to operate a proper system for passing on pilot landing positions to crews beyond radio contact. As a result one crew drove 900 km to Skopje and back when their pilot had landed only 30 km out. Each time they had telephoned they had been told 'no news'. This was on the second day's task and far to the south of the country not only did roads and telephones possess character-building elements, but language became an increasing problem; perhaps as well when translated road signs said things such as 'Beware of Wolves'. In one field five pilots landed to be followed rapidly by police who refused to let them talk to each other. Because most gliders flew much the same distance on this day the scores of the thirty top pilots were close, the real contest having been for the trailers which had clocked up 60,000 km by the time they returned a day later, some of them damaged. Then the rain started and went on for four days. The trailer park was flooded and my Jury office floor in the new ziggurat-shaped control tower drained from the balcony to my desk, so I worked in wellies. The 5 tons of water ballast supplied each day for the gliders became lost in the widening airfield lake. Then there was one difficult task-day before the thunderstorms came; these were no ordinary storms, and they brought disaster. Lajos Varkosi of Hungary was killed by a direct lightning strike at about 9000 ft near Pancevo, and he crashed with his glider, parachute unused, into an area where the storm was blowing down roofs and trees. Only one pilot succeeded in completing the task and returning to base, Nick Goodhart, with his electrics blown. Other pilots suffered serious hail damage, or were tossed from 6000 ft to 600 ft at 8 m/sec, catching a mere glimpse of nearby ground before being swept up again; and few did not say openly how frightened they had been. The storms continued and the organisers continued to send everyone off into them. Now the violence in the air became less extreme, but the ground had become impossible, with gliders, trailers and cars stuck in mud. Twice the British team had to use three Range Rovers on an exhausting and muddy overnight retrieve to collect one glider. They were also used to help other crews who had ordinary cars – the Range Rover attached astern of the trailer being towed to hold it from sliding sideways into even muddier ditches. Competitors helped each other, with their cars, their radios and with their muscles.

But this Championships had not yet finished with its pilots. On the penultimate day my Canadian friend, Wolf Mix, was approaching low into a field but touched the road, and swung round into the path of a truck which had just squealed to a stop to try to avoid him. A helicopter took him to the best neuro hospital in Belgrade. It was an accident which could have happened in any Championships in any country, and I felt sorry for the organisers that this had come on top of all their other troubles. That evening a call came from the hospital for Wolf's special blood group, a public radio appeal having already produced volunteers, but not all were suitable. Running to the dining-room where several hundred crews and helpers were eating, I called for anyone who had this special group; and was amazed. Without a second's hesitation five people just put down their knives and forks and came over; a Norwegian, a New Zealander and the two children of the Italian pilot, Pronzati, brought by their mother. Outside a helicopter waited to fly them to the hospital; but it was no good, and Wolf died early on the morning of the last flying day.

Again the forecast gave cu-nbs with low cloudbase and heavy rain, and again a task was set – a 238-km triangle for both classes. This was asking for even more trouble but most competitors were by now too tired to complain. At least the two classes should have been given different tasks as recommended by the CIVV; not only to let each get on with its own contest, but to reduce the risk of collision. I and others had already made this point to the organisers, but the significance was obviously not getting through. So launching started, the gliders disappeared, and thunderheads towered into the sky. At base everyone waited, latching on to any snippet of news from any radio, so talk of a collision spread like the floods still around us, with great relief for anyone who knew it wasn't their own pilot. It was Wroblewski who had called his crew to say that he had seen two gliders fall out of cloud near Belgrade. That was all. Then Sheila Innes phoned in to say that David, now flying for Guernsey in the Open Class, had collided with Ake Petterson, the Swedish Standard Class pilot. Both had parachuted, Ake spraining an ankle and David breaking one – and having to be taken to hospital by boat through the floods. Both gliders were wrecked but had not fallen on anybody. Several other gliders were damaged on landing, A.J. Smith cutting his knees on the instrument panel and then having his turn-point cameras stolen, dropping him from sixth to sixteenth.

After eighteen years without a fatality, Vrsac was a sad Championships because there was much that was good, the launching and the start line observers, the food, the hospitality, and the high hopes, but by the end everyone was just glad to be able to go home.

It is events like this Championships which breeds thoughts of whether what one is doing is worth while, which in turn leads to thoughts that maybe there are other things in the world as well as gliding that should be enjoyed before one becomes too old and it is all too late. Back at Farnham I found Lorne in the midst of deciding that it was now his turn to buy a boat. Little *Sula* had given us so much fun but now we both wanted to cruise further than was possible – or comfortable – with the Nimrod. We still wanted a drop keel boat so that we could explore creeks and rivers as well as the sea and after a great deal of research decided on a Dutch-designed 22-ft Etap, which almost exactly fitted our specification. With regret we sold the Nimrod to Frank Irving, and the Etap we also called *Sula*. As with any new boat – or glider – there was a lot to do to get her as we wanted, but with a trailer boat the work could be done at home. By Easter we were on the water. We gave *Sula* a bright spinnaker, raced her with the Bosham Sailing Club – being pleased to sometimes win – and found our gliding and flying useful in making the most of wind and weather. That summer for our holiday we sailed *Sula* to Cornwall, discovering the delights of leaving a still sleeping harbour at dawn and watching the sun climb out of a pearl-coloured sea; or racing for shelter with the rising wind tearing off the tops of the waves, and then on a mooring enjoying the sudden quiet and snugness of the cabin. We even had a few days idling over a summer sea, sails just filling, hearing the cry of gulls.

To me being on the water was like home, and I realised that over the years I had not been there often enough. So that November I went to sea again, as a foredeck hand on the three-masted schooner, *Sir Winston Churchill*. (I think I was their first grandmother.) Sailing ships had always fascinated me ever since, at sixteen, I saw the great grain ship *Herzogen Cecilie* sail across Start Bay, the evening sun gold on her royals and topsails; shortly before she died on the rocky coast of south Devon and at low water I clambered across her jumbled deck. Being on *Sir Winston Churchill* was a delight. In wild winds, or moving quietly through the night on watch with half the crew asleep, was age-old magic.

18 · Honorary temporary Australian

All my life I had been attracted by Australia but never had the chance to go there. Suddenly there came the riches of two visits in one year. This good fortune had come about because Kenneth Davies, who for many years had been the Royal Aero Club delegate to the FAI, wanted to retire and proposed that I take over from him. The Royal Aero Club itself was also going through a big change, as for years, since it left its own home at 119 Piccadilly, it had existed in a London Clubland wilderness, increasingly removed from flying. Philip had been pressing for some time to reverse this trend and, on becoming Royal Aero Club Chairman, set up a committee to decide how to best go forward. This, together with my now becoming the FAI delegate, brought me into closer contact with all the other special cells of sporting flying. I had at one time or another tried ballooning, parascending and aerobatics, so it was easy for me to believe, as Philip Wills did, that we should all work together if our sort of aviation was to survive the increasingly heavy hands of bureaucracy, and inflation.

My first official FAI Conference had been the previous year in Dublin as Observer with Kenneth; now I would be going to Sydney in November as Delegate. But before that, and missing the dreary February of an English winter, there was the 1974 World Gliding Championships on the edge of the South Australian desert at Waikerie, where I was to look after the International Jury again. I could hardly wait to go.

Qantas flew me in luxury across a new and beautiful world, and within hours of arriving at Sydney I was sailing on the Hawksbury with friends of glider pilot Tommy Thompson, the Qantas doctor. Next day I set off with Sel Owen to drive the 600 miles to Waikerie under a hot cloudless sky, cold winter forgotten.

We arrived to find a superb organisation created by Wally the Met, who had emigrated with the Lasham anticyclone, and was now Director of the Championships. The place was full of gliding helpers from all over Australia, whose enthusiasm and sense of fun was highly infectious; and I had been given a caravan to live in on the airfield, which was ideal. The only problem was the weather. Obviously the immigrant anticyclone, afraid of being overworked, had gone into hiding. It was unbelievable. For several years the local gliding club had lovingly cultivated two patches of grass on the desert airfield, so that competitors would not have to take off in a dust fog. Then, a week before the Championships it rained for the first time in years: and it did not stop. We arrived to find the much-publicised grass patches lost in lush green to the horizon, with caravans and garden sheds floating down the flooded Murray River. As someone said, 'If you want rain in the Sahara, just organise a Championships.' Others just remembered Damphill and, more recently, the ill-fated storms of Vrsac.

During the practice week the floods grew, upsetting several well-laid plans including the collection of two Murray River houseboats for competitor accommodation, from Renmark, 117 miles upstream. One was for the New Zealanders, who were prepared to collect it themselves, while four young Australian glider pilots volunteered to fetch the other one. Gordon Hookings, an old New Zealand friend of many Championships, and I went with them, being made temporary honorary Australians for the voyage, all of us arriving at Renmark very late in the evening. The houseboats were 40-ft long catamaran hulls on which a luxurious holiday chalet had been built, with two rooms, kitchen and loo, and we were busy stowing toothbrushes and beer when the owner, not unreasonably, demanded to know what experience we had in driving his valuable property on a flooded river; totalled, it came to the youngest Australian having once been a houseboat passenger for half-an-hour. So, still not unreasonably, he demanded that at least one of us should take the boat out into the dark fast-flowing river, under his watchful eye, and bring it back to its berth. 'You,' he eyed me firmly, and started the engine which drove the huge stern paddle-wheel. With my honorary Australian status at stake there was nothing for it, except to imagine I was going to do a circuit on a Tiger Moth in a gale: it must have been about right as I managed to bring this monumental vessel back to its berth without vanishing downstream into the night.

Inevitably this trip became a contest between Aussies and Kiwis, who tried to steal away before first light, only to fetch up stuck on the next weir downstream. We were hailed from the bank in the middle of our breakfast by the owner who wanted to use us to ferry his engineer to the stranded houseboat. He also used us to try to pull them off the weir, but they remained firmly glued to the concrete until a powerful river work boat got a grip on the situation. A little before 8 a.m. we were released and fled downsteam, leaving the Kiwis against the bank occupied with some final repairs. We now swirled along at a goodly speed, made even faster when the Australians discovered that the throttle could be held more than usually wide open by cunning use of a wire coat hanger; and having done this opened the beer. And so we swept down this rushing, muddy river which now encompassed hundreds of eucalyptus trees on either bank, as though we were charging through a watery forest – navigating by little sketch maps in the brochure. The coat hanger was holding up well, giving us an extra knot or so but at the expense of some hydro-dynamic buffeting which rattled the beer in the fridge to such an extent that fears were expressed for its safety – with the obvious remedy. Suddenly the Kiwis called us on the radio – we had taken glider transmitters with us – hoping to hear that we had broken down or run into a tree, but on discovering how far we had gone lapsed into silence.

A beer count showed that it would be finished by 1630 with our earliest ETA 2200 hours, so to divert attention from this potential disaster genuine Aussie Davy Jones took a ride in our towed aluminium dinghy which, at the speed we were going, promptly developed violent instability in pitch. To avoid going straight to the bottom followed by ribald remarks about his name, he baled out.

I was driving the houseboat at the time at the front end and the first thing I knew of the goings-on at the back was when the remaining beer stopped rattling in the fridge as the revs dropped. I was just wondering which idiot had removed the coat hanger when '*Man overboard!*' was shouted: the idiot had intelligently restored to me control of the throttle as first action. Turning back I found it just possible to stay stationary against the current, which was much better than having to try picking someone up while navigating downstream backwards. By this time all four of the genuine Aussies were in the water, splashing and swimming around like hilarious seals. It was only when Gordon and I said loudly that if they did not come back on board the Kiwis would beat us that brought them reluctantly out of the water, and to replace the coat hanger. For the rest of the afternoon we careered down this fascinating river, at one time through floods almost to the horizon, until darkness fell. Without spotlights all hands watched for floating debris with such concentration that when we finally swept into Waikerie town at 2215 – and managed to stop – there were still two bottles of beer unopened in the fridge. 117 miles, and one and a half hours ahead of the Kiwis.

After a few more practice days the Championships' opening ceremony approached with everyone, as usual, longing for it to be over so the real flying could start. Prime Minister Gough Whitlam came to do the honours but as Pirat could not come I was stand-in for him on the platform. It was typical: the wind blew so strongly that we were in severe danger of being blown off the dais – unless it took off first – while waiting for a low-flying glider to symbolise the Opening. It came past us fast, only a few feet up, with the spectacular trail of discharging water ballast exactly as planned. Prime Minister Whitlam, the Premier of South Australia, Don Dunstan, and I could not have been better placed to receive it. Such is the price of fame. But next day the rain stopped and with remarkable rapidity the temperature went up and the river went down.

Sixty-seven competitors came to Waikerie. The South African pilots were not allowed in the country but sent a good luck telegram which was cheered when read out at briefing, but apart from this many faces were familiar, including Hans Nietlispach, the Swiss pilot who had first competed at Samedan in 1948, and Dick Johnson from Spain in 1952. Italy's Adele Orsi was the only woman competitor. The British Team had brought George Burton and John Delafield for the Open Class, with Bernard Fitchett and John Willy in the Standard.

As soon as the weather improved the daily *Championships Bulletin*, run by Martin Simons, could report real flying, instead of filling the pages with hints on dealing with snake and spider bites, escaping from bush fires – by setting fire to the grass downwind and then running after it in thick shoes, avoiding sunburn or salt loss, and the need to carry two litres of survival water – to drink after landing in the floods. It was all useful, and further backed up by real help from the Australian Post Office which had produced a list of the whereabouts of all the remote telephones in the thousands of empty square miles over which tasks might be flown. This also helped pilots to avoid landing by homesteads which were deserted, as Norbert had done at Marfa: though one Waikerie local added that pilots should land only at homesteads where the piles of bottles outside the back door still glinted cleanly in the sun!

So the Championships flying started with all its tense excitement, and a flurry of activity within the organisation to replace orange ground marks with white ones: pilots were being confused by farmers laying out their apricot harvest to dry in the sun. Mostly the tasks were 300- and 500-km triangles, successfully completed in the strong thermals by the lean white gliders, but near the end of the two weeks the Open Class was given the largest triangle ever in any contest: 707 km, with turn points at Clare and Nangiloc. Departing around noon it would need 7–9 hours' flying to complete, with, for part of the flight, competitors being more than 300 km from base over empty country; so it was only sense to put up a Waikerie Air. I went in the air-conditioned DCA Bonanza – a pleasant change from the 38°C on the ground – and we flew the route, checking any gliders we saw far from habitation, and passing the usual crew messages. In the meantime the Open Class pilots were fighting on, working around a great chunk of frontal cu-nb that had developed, and which from Waikerie appeared to lie right across the way home. So it was with delighted surprise that Marfa champion George Moffat in his Nimbus II and François Ragot in the ASW–17 crossed the finish line together at 9 p.m. Soon came eight more – Holighaus and Grosse of Germany; Ax, Sweden; Cartry, France; Zegels, Belgium; Delafield, Britain, and Australian Tony Tabart. They were the lucky – as well as the clever – ones. Some of the others would pass another desert night cramped in the cockpit, wondering when the trailer would come. The sun was setting the big cloud on fire, as we flew towards home looking for a lost trailer; but it was dark when we found it, as at Marfa by its lights in a smooth sea of darkness. It was getting on for midnight when we touched down at Waikerie along a line of paraffin flares, to join the celebrations in the bar.

By 23 January there had been so much flying that everyone was really tired, so when the weather looked like developing late a rest day was declared to everyone's relief – most pilots just slept – while the weather grew even better. Perhaps the most exciting task of all was the 509-km triangle at the end of the Championships, because it was completed by every pilot in the Open Class and twenty-nine out of thirty-eight in the Standard. Not only was the flying good – Goran Ax won at an average speed of 140 km/h – but the evening which followed with everyone together was a nice change from the long retrieves. These, for some pilots, were becoming not a little monotonous and the road signs – 'Wombats for next 7 miles' – had long since ceased to raise a laugh. Then after one more task it was suddenly the end.

Moffat had for some time been a certainty for Open Class champion and he beat Bert Zegels by 400 points and Hans Werner Grosse by almost 600. In the Standard Class Reichmann did not have such an easy time, beating Australian Ingo Renner by only 29 points and Kepka, Poland, by 59. Bernard Fitchett stayed in third place until the last day. Then once more the rain came down, flooding the airfield, and the wind blew so noisily that the prize-giving and the speeches were inaudible. But no one minded any more. It had been a wonderful Championships and a safe one, and the only glider damaged beyond immediate repair was Canadian John Firth's Libelle. Landing in a paddock he had to ground loop to avoid a trailer, driven by a tunnel-visioned driver hurrying to another glider.

Even on the Jury there were no serious problems. Hans Werner Grosse was not allowed to change to alternate wing tips, there was an unsuccessful objection against the Directors' decision to reduce the score of several pilots by two points for small technical infringements, and there was a complaint against a pilot who dumped his water ballast in a thermal spoiling the laminar flow of gliders circling below. This would be yet another fascinating subject for the next CIVV meeting.

I left Waikerie with the Wallingtons to stay in Canberra, where their son John tried to teach me to roll a canoe, but I kept laughing every time I was upside down so it was not all that successful. A few days later we drove over the blue mountains to a friend's beach cottage near Bateman's Bay, bathing in gigantic Pacific surf early next morning with a kangaroo the only spectator on the beach. Then to Sydney and by 747 over the Pacific, taking photographs from 30,000 ft of cumulus reflections in the quiet sea, and on to Fiji, Acapulco, Mexico, Nassau and Bermuda gazing at the world and only sad when it was dark over the Caribbean – though some compensation was looking at the flickering trail of the Florida Keys on the flight deck radar. Heathrow Airport at six o'clock on a cold, windy morning was an anticlimax: it was good to be back home, but I was already looking forward to returning to Australia.

It was soon after the October CIVV meeting with its

expected long discussions about dropping water ballast on other gliders; or carrying so much ballast that the certificate of airworthiness was invalidated; or whether there should now be two competition Classes or three, that I left again for Australia. This time I flew British Airways, to check in at the Wentworth Hotel in Sydney with the other FAI delegates for the Conference.

As usual the opening ceremony, which took place in the splendour of the Sydney Opera House, was combined with the presentation of annual awards. This was special for me, as I was to receive in this building I had so long admired, but never expected to visit, the FAI Lilienthal Medal for gliding. Afterwards all two hundred of us delegates and friends were to cross the harbour in a large ferry boat to the residence of the Governor General, Sir John Kerr, who was giving us lunch. On the way over I asked the helmsman in his glass box about the boat's steering, remembering a similar sized ferry off Sardinia on which the helmsman was over eighty because the steering was so awful that no one else could learn to manage it. The Australian grinned. 'Try it,' he said, and stepped away from the wheel. This wasn't what I had meant, but it was also an opportunity not to be missed. So I took the wheel with some trepidation, but quickly found the steering positive and straightforward, while my friends made funny faces through the glass. All too soon we were approaching the ridiculously small jetty at the bottom of the Governor General's garden – it looked about right for one of Queen Victoria's lesser launches. So I lifted my hands from the wheel and offered the helmsman his ship back. 'Have a go mate,' he said. This was even less like any of my intentions. I had never driven so big a vessel, nor taken any boat into such a diminutive jetty. It looked impossible, but was also slightly ludicrous producing a strong inclination to giggle. On the sloping lawn the smart uniformed band was playing classical music, various flunkeys in white gloves were waiting to escort us, and on the jetty itself was an extremely smart officer resplendent in white and gold uniform to greet us. I had already reduced power but insisted that the proper helmsman continue to deal with the throttles – never try to learn more than one new thing at a time. Studying the tiny jetty with its three little green bollards – only fit to tie up a small and ancient steam launch – I decided on a glider approach pattern and very slowly indeed came alongside, the real helmsman doing some clever work with the power. The problem was that our giant ferry had a wide fender all around it for dealing with normal harbour walls: I reckoned that this might be higher than this little jetty. It was. As we all trooped down the gangway to our reception there was the white and gold officer still standing smartly to attention, but now with a little green bollard in his arms, held like a favourite baby. 'That's all right mate,' said the real helmsman, 'it had dry rot anyway – it's only the paint that's new.' Lunch was marvellous. After the Conference I had two lovely weeks staying with friends, and there was little inclination to return to the English winter, even though I would be going home via Tahiti, Hawaii and Los Angeles, with a stopover there to stay with Lloyd and Rosemarie Licher, whom I had not seen since Marfa.

As the 747 again took me out over the Pacific with Sydney Harbour and Botany Bay fading into the haze, I settled down to enjoy the rest of this second circumnavigation. By now I had discovered that jet lag seemed no more of a problem than a slight tendency to doze off in the middle of an important conversation the following day, and that flying eastbound around the world was marvellous as there were more days and shorter nights – more to see; meals came round regularly with someone else doing the cooking and washing up; and being lucky enough to have been provided with a first-class ticket, the voyage was not a minute too long. There was not only the world to look at, but the world's weather to photograph – inter-tropical fronts, cu-nbs, sea-breeze convergence zones, it was all there for the taking.

In Los Angeles I met many American hang glider pilots, saw their new aircraft, their flying sites, and how they got on with conventional gliding. Here, too, a closing of the ranks was apparent against this seemingly maverick aviation; but much less than in Britain. More conventional aircraft designers and pilots were involved, and the attitude I felt was nearer that of indulgence than any 'it ought to be banned' attitiude. Paul MacCready, 1956 gliding champion and later of Gossamer Albatross fame, and his sons were all into hang gliding, and it was good to listen to refreshing argument. In California it was just a great getting together of enthusiasts, full of new ideas and theories, but no more than in Britain was it yet settling into any long-term direction. Free flight was the cry, a revolt against the concrete and complexity of sophisticated aviation for most, and against conventional society for a few. The visit to Disneyland with Rose Marie seemed less of a fantasy – though just as much fun.

When I got home to Farnham I found moves to amalgamate the National Hang Gliding Association and the British Kite Soaring Society in turbulent progress, with a meeting arranged for 8 December 1974 at Coventry. Would I chair it? Presumably as the interested person least directly involved in the problems. It was obviously sensible to have a single national association, but any amalgamation is always riddled with difficulties because inevitably some people have to give up positions of power which they may well have created by hard pioneering work, and in anything new, such as hang gliding, people with ideas are more plentiful than those prepared to run a routine administration. If I was to take on this meeting I needed to do some homework. This started with a long talk with Bob Mackay and Dick Bickel, who represented the two Associations, and then sitting down with lots of paper to set out what we should aim to achieve, and how to progress the meeting so as to obtain the objectives. One certainty was that many people were going to want to have their say, and must be allowed to do so without the meeting getting out of hand. So 8 December arrived, and so did I and some 200 enthusiasts at a cold and dreary hall just outside Coventry. Not unexpectedly the argument and the discussion and the thinking aloud went on all day, but through it all, though not always apparent, was a sense of common purpose, and after seven hours of it we had a new Association – the British Hang Gliding Association. We also had a new Chairman, with some reluctance, as no one seemed keen to take on this responsibility; but finally Martin Hunt gallantly stuck his neck out. We had a fine Secretary in Chris Corston, and an individualistic Editor for a new little magazine called *Wings!* in Nick Regan. BHGA was in business. Now it had to get down to the real work of reducing unnecessary accidents, devising pilot proficiency standards, producing instructors, sorting out airworthiness problems, and learning to live with other pilots in the same air. Everything, in fact, that years back we had done in gliding.

It was going to be a long and turbulent journey, with mistakes made or repeated along the way, but too many people wanted this sort of flying for it to fail. Someone was even heard muttering about an idea for a World Championships for hang gliders.

19 · Kössen and Canada

It did not take long for the rumours about a World Hang Gliding Championships to spread, though the FAI in Paris had no more information than anyone else. It could not be an official event, because no FAI rules existed as there was yet no Committee to write them, but the excited whisper persisted that some little village in Alpine Austria was going to run a world competition – whether it knew how to do so or not. I had taken little notice of all this until I got a telephone call from Ken Messenger, who had heard from someone in Austria that they needed a jury member from Britain. Would I be interested? Having long since learnt that the best opportunities only come once I immediately wrote to the address Ken had given me, saying that I would be willing to go, and asked without much expectation of success for copies of the usual paperwork – rules, entry lists, and so on. All that came back was a nice letter of welcome saying that I would have accommodation at the Haus Pirmöser.

The dates coincided with the March CIVV meeting in Paris, so after it finished Asahi Miyahara, the Japanese delegate, and I got on a train to Austria. Asahi had been involved in aviation even longer than I had, and we had much in common as our concern for the well-being of the little end of flying was the same. We arrived at Kössen at the end of winter. The skiers had wiped the muddy snow from their boots and gone home, and the alpenblumen watchers with their walking sticks and haversacks had not yet arrived. So in between came the hang glider pilots – over 300 of them from twenty-two countries when only fifty or so had been expected. It was a world pilgrimage and the 2,500 villagers were delighted, writing a special fanfare to be played by their brass band, and setting up an office to cope with everything from finding accommodation to writing some rules. Over it presided a tall and cheerful Sepp Himberger, wondering how to run a competition of 300 pilots whose skill and experience were unknown, and whose only thought was to get airborne. Forty-four of these pilots had turned up from Britain – just about everyone who owned a hang glider, fifty-six came from West Germany, forty-nine from Switzerland and twenty-five from the USA. Sepp had planned a take-off place on the Unterberg mountain for either foot or ski launching, with landings on a small field in the valley 600 m below, which was gaily decked with national flags. Their poles looked an interesting hazard until glider performance proved so poor that these tough sticks could not be reached; some gliders being in ground effect all the way down the tree-covered face of the mountain. But no one was bothered except Sepp, who still had to reduce 300 individuals to a manageable number of competitors if the contest was to work. A preliminary target-landing event got the elimination process going well, as some pilots had never flown from a height of 600 m before and had to concentrate so hard on getting down safely that they missed the 50-m target entirely: others had hang gliders that would only go straight, like paper darts, and more of these missed the target as well. But the enthusiasm, the sense of fun, and the furious arguments about this new flying were great.

By the official starting day we were down to ninety-seven eligible competitors, the others being allowed to fly after the daily tasks were finished. We, the jury, observed the landing judges tirelessly endeavouring to measure the distance from target centre to the pilot's foot while running in all directions to avoid being impaled by the next great swooping dart. The seven of us, including Dan Poynter from the USA and Bill Moyes from Australia, had been provided with a smart raised box from which to see, complete with windbreak and girls in national costume who supplied us at frequent intervals with tots of fiery liquid to keep out the cold. Each day we watched the noisy rustling wings either from our box or from the take-off point up on the mountain, and each evening talked hang gliding over steins of beer – what would be the future, or of more immediate importance, what constituted a 'good' landing: should the distance from the target centre be marked from the first footprint or where the pilot stopped? One thing we did know was that the first footprint assessment was resulting in some remarkable kamikaze dives for the spot at which the pilot would kick with a foot as he screamed past before fetching up in a heap somewhere else.

It is always a pity that the enjoyable happenings in life finish so quickly. All too soon the fifty-strong village band, with its age range from eight to eighty, played their fanfare for the last time and quantities of prizes were presented to winner Dave Cronk, from the USA, with his Cumulus, to the runner-up Werner Scherne from Switzerland, on an American Phoenix, to eighteen-year-old Roy Haggard, also from the USA, on a Dragonfly, and to many other competitors. This first experimental competition at Kössen had been a safe one without any serious accident, and everyone wanted to come back again. Sepp had begun to get the world together, and now the FAI needed to give this new simple flying its blessing and help it on its way.

Later that spring I was invited to Paris by Bernard Duperier, then FAI President, to talk about the possibility of introducing a new International Committee, the CIVV having decided that it did not wish to incorporate hang gliding because it had enough work with conventional sailplanes and motor gliders. Within a few hours Bernard decided to recommend to the FAI Council that the Committee should be introduced. The Council, at its June session, agreed and I was asked to chair the inaugural meeting of delegates who would be nominated by their

Kössen and Canada

national Aero Clubs. Now came a problem, as some knew little about the hang gliding enthusiasts in their country, nor how to contact them, and many hang glider pilots had never heard of their Aero Club. Nevertheless, eighteen delegates from thirteen countries turned up and we immediately started work on pilot proficiency standards, record categories and, most urgent of all, championship rules. Sepp wanted to hold the first official FAI Championships at Kössen in September 1976, and rules not only had to be drafted, but approved by the FAI and circulated six months before the event. What was remarkable was that the collection of individuals who came together at such short notice from all over the world succeeded in keeping to this demanding timetable.

So, within fifteen months of the FAI setting up its international Committee, CIVL (Commission Internationale de Vol Libre) we were back at Kössen with a full official World Championships to play with. But this time competitors were entered formally by their own countries, so at least something was known about whether or not a pilot could fly! Numbers, too, were more manageable with a mere 150 pilots from twenty-six countries. But tasks were still mainly for landing accuracy, though with a few more turnpoint pylons to match improving glider performance. September had been chosen to give time to write and circulate the rules; and it was also between seasons, with the summer tourists departing before the first snows. But as the minibuses and cars arrived, covered in rolled hang gliders and filled with smiling pilots and all the miscellaneous junk that seems an essential part of any aeronautical competition, the enthusiasm and the longing to fly was the same. And the flying was good because the newer hang gliders enabled pilots to use thermals – if they could find them, so it was the same sort of exciting, pioneering flying that had been so marvellous in the early days of gliding. It was also without accident to any competitor. The winners, even younger than 1975, were the Australian Ken Battle, who at eighteen won Class III; Terry Delore, seventeen, became Class II World Champion for New Zealand – he had started flying aged ten – with runner-up nineteen-year-old Dean Kupchanko from Canada. The

Standard Class was won by the Austrian Steinbach brothers, who had developed a remarkable target-landing technique by virtually stabbing the spot fully stalled with the rear end of their keel tubes. Sixteen-year-old Australian Ricky Duncan was in third place.

After most of the competitors left for home we had the second CIVL meeting to work out how hang gliding could best be encouraged and helped internationally. As its first and founding President I had said at the inaugural meeting that as soon as the delegates had got to know each other well enough to hold elections I would stand down. I knew that it was essential that hang gliding should be developed by people who were actively involved and responsible for their own actions. This brought American Dan Poynter to the top of the pile, with myself voted into the usual retirement office of Honorary President. I could still help them, but the decisions, right or wrong, would be theirs.

Kössen was beginning to become a tradition, a Mecca to which everyone wanted to return, so there was delight when Sepp Himberger said that he would run the FAI European Championships for 1978; and to allow everyone to go to Kössen, it would be open to the world except for the title of European Champion. So once, again, in September, we all met together: 135 Europeans, plus familiar faces from Brazil, Canada, the USA, Australia and New Zealand; with myself looking after the International Jury. Only the gliders were new: sleeker, faster, and more manoeuvrable, capable of soaring if the weather was right; though in the best international tradition it rained heavily for the first four days and the contest ended the same way. But in the middle we had the first ever cross-country task: to fly as far as possible along a high-sided Alpine valley. The distances flown were insignificant. Pilots had not yet learnt the techniques and cunning of the glider pilot and it would take time for them to learn to explore this wide new sky. But delight was total. This was real free flying at last, and the end of repetitive target landings on the home field: from now on landings would be in new, strange places, as far away as possible. I went in one of the patrol cars to check landing positions and every time we trekked across a field it was to

Kössen and Canada

find a pilot exploding with excitement and happiness. The dream of flying like the birds was becoming as near reality as it would ever be. I knew just how they felt: it was the same delight of my early flying years that I would love to have again.

These Championships, too, were safe, the only incident occurring during evening free flying when a young pilot on a borrowed glider became over-exuberant and whip-stalled. The poor over-stressed aircraft collapsed and fell a hundred feet on to the ski lift cables – with the pilot suspended unhurt between them!

Now that soaring cross country was effective as a competition task there were great hopes for long flights at the next World Championships. These were to be held in July 1979, on the great white limestone cliffs above Grenoble in France; a superb backdrop for 181 bright butterflies dancing in the hot sun. As International Jury Chairman I lived with the organisers in the valley, which caused me to rapidly brush up such French as I could recall, but our packed lunch was brought to us at the launch point up on the mountain. Each day flying was almost brought to a halt for this sacred gallic midday hour while we consumed quantities of bread, cheese, salad, fruit, and wine.

We had a few protests, but not serious ones, until the last day when we had five, all concerning the number of turns pilots had, or had not, made around the pylons. The organisers, who had arranged a fine farewell party in the cellars of a local chateau, were all for dismissing anything remotely like a protest meeting – for obvious reasons – since whatever was decided the top pilots would still win. But to ensure that all objections were dealt with fairly was what I was there for, and I knew that even a pilot from a small country placed somewhere in the middle of the list instead of the bottom felt that his true placing was important. Inevitably, it was late by the time we had worked our way through the protests; four in which the organisers' decision was upheld, and one in which the protester gained his claimed score. As soon as possible the Jury – myself plus one member from each country – dashed up the mountain before the contents of the cellars were finally consumed. We just made it – and heard only the last of the speeches in the stifling heat of the packed underground room. This was made by Henri Fabre, whom I had met a few times on the landing field. A great supporter of hang gliding, he was ninety-seven, drove his own car, and was the inventor of the seaplane in 1910! It was a great spanning of history for the hang glider winners to talk to someone who had earnestly discussed flying machines with Wilbur Wright.

The excellent weather at Grenoble had given thirteen consecutive contest days and over 3000 flights, once again without anything passing as a worthwhile accident. Gerard Thevenot of France won, with Britain's Johnny Carr second, but although no long-distance tasks were set, both gliders and pilots were now more than capable of serious soaring. In the USA George Worthington had gained the world distance record of 153.61 km, which was also a goal flight, while even England's temperamental weather had given Rob Bailey 80 km. The BHGA, under the chairmanship of Reggie Spooner and then Roy Hill, had grown into a well-respected organisation whose members had gained seven of the first ten international Delta Silver badges, which had the same 50 km distance requirement as conventional gliding's Silver C. Yet it was only five years since these 'ragwings' had been looked on as a somewhat dangerous joke by many; and it seemed no time at all since the day Lorne and I had gone out on the chalk downs near Marlborough with Geoff McBroom and his home-made hang glider. It was about the first in Britain, an enormous blue and red floppy wing, uncontrollable except by Geoff. The day was perfection, with cumulus shadows drifting over the rounded hills, and everywhere thousands of wild flowers. We rigged his great sail among them and when Geoff floated into the air, suspended in sunshine, I was filled with excitement. Suddenly I remembered soaring on this same hill one glowing evening forty years before; six of us in slow, light gliders, the wind blowing our hair and one pilot singing as we floated on the quietly rising air.

The 1975 FAI Conference was in Canada, and it was there I had the chance to fly a floatplane; something I had always

Kössen and Canada

wanted to do. Bob Purves was one of the chief Conference organisers and he and Doreen asked me to stay with them in Winnipeg after it was over. Doreen left Ottawa early to get things ready, leaving Bob and me to follow by jet as soon as the hectic packing up was done. We just managed to get to the airport in time, to find the aeroplane delayed by an hour. This could be difficult as the plan was, on arrival at Winnipeg, to jump into Bob's Cessna 180 amphibian and fly to their summer house at Lac du Bonnet, about 40 minutes away, before dark. At long last the jet took off, flying west towards the lowering sun. Bob asked the stewardess to find out from the captain the time of sunset in Winnipeg, but the reply came back that he did not know – presumably being of no importance to him. So Bob got out his HP65 calculator and started to work it out, while I looked at the angle of the shadows on the cloud tops, made my own assessment, and came up with anytime between seven and ten past. Bob's calculations gave him five past. In the event we landed at Winnipeg at four minutes past seven just as the sun slipped its red disc below the sharp horizon! Bob's son was waiting with the car to drive us fast to the Cessna; only to find that it had not been refuelled as requested – one more delay. In the fading light Bob pre-flighted the aircraft while the fuel bowser arrived, and my deep-rooted caution came to the surface. This was exactly how accidents happened: rush, hurry, something forgotten, and a pilot fatigued from a week of serious overwork – and running out of daylight. So I watched carefully, doing a little back-up monitoring. But I need not have worried, Bob's checking was meticulous.

Airborne at last off the long runway of the international airport, it was obvious that we would have to land on the lake in the dark, without any illumination except randomly from weekend cottages on the shore. Bob said he had done it before, and as a pilot I was beginning to have great faith in Bob. He would let down, he said, at precisely 200 fpm at precisely 65 mph airspeed, and in due course we would arrive gently on the invisible blackness that was the surface. It was a technique developed for landings on glassy water and unmarked snow. Over the end of the lake Bob started his let-down and we sat there seeing nothing but the dimly glowing instruments. Then the airspeed slowed just a little below the critical speed and Bob lowered the nose fractionally to correct. But this would mean that, without further corrections, we could arrive on the surface in an attitude which might cause us to nose over. No dice. Without hesitation Bob decided that tiredness was taking the edge off his concentration, abandoned the approach, opened up, and returned to normal flight.

'We'll go into the strip,' he said, turning to fly over some lights on the lake shore. 'Doreen will fetch us in the car.' At a few hundred feet he blipped the throttle. 'That's our collection signal, now, here's another one.' He switched the radio to 122.5 and pressed the transmit button three times. It was magic. Out of the darkness almost directly ahead a fully lighted airstrip sprang from the night. 'All the houses on the lake with aeroplanes can do this.' I looked at Bob, who was smiling with the fun of this playing in the air. Round we came on the approach, water rudders up and wheels down – important to have it the right way round – and landed gently on a narrow strip cleared in the forest with its end at the lake shore.

The next day Bob and I flew over the lake country to the east of Lake Winnipeg. The excuse was to check that a fishing dinghy belonging to his family was still at the little island on which it had been left back in the summer, but the real reason was just to fly. Map reading was fascinating – there were thousands of lakes, their still surfaces reflecting the blue of the sky and the brilliant fall colours of the forests. Of habitation there was no sign, nor of people. You needed to mistake only one lake for another and finding yourself again could be life's biggest problem. Bob knew this country so we were in no risk of wandering, but to me the challenge of not getting lost was real and concentrated hard work, but success felt good when the right funny-shaped piece of water turned up in the right place. After a couple of hours flying we reached a large irregular lake studded with islands, and prepared to let down. There was almost no wind to roughen the surface, but various reflections helped to give reference. Gently we touched, and taxied over the pale smooth water to one of the islands, the ripples of our

Kössen and Canada

passing spreading wide behind us. Running gently on to a little beach we climbed down on to the floats, and from there to the sand to turn the Cessna around, and tie its tail to a tree – and listen to silence. We were probably the only humans for 50 miles in any direction, and for me this was what aeroplanes were for: not the noise and sweat of airport flying, all concrete and chatter. It was exploring again; of course people had been here before, but there was no sign – and we could have been the first.

As the ripples died away with quiet whispers when they splashed the shore, we went to look for the dinghy among the trees. It was untouched, and as we turned it over, there was the outboard underneath and the cans of fuel. 'We can have our lunch on the Lake: it'll be good to run the engine for a bit.' Bob smiled. It was not difficult to drag the aluminium dinghy over the few yards of beach, and the engine started at the first pull, its noise reverberating back from the forest, raising an unheard protest from a nearby foraging duck. Speedily we went out past the end of our little island and over the smooth water towards another even smaller one about a mile away. As we neared the rocky outcrop of its shore Bob stopped the engine; and the silence hit us again. We ate our lunch gently drifting with just enough movement to change perspective and reflection, and listening to the mournful call of a far distant loon. It was one of those moments in time that I wanted never to end. But it did end, with the usual practicalities of life – we had finished our lunch and we had promised Doreen to be home before dark. So with one more shattering, noisy rush of engine we returned to the Cessna, pulled the dinghy back into the trees, turned it over, and climbed back into the aeroplane. It seemed like home; and it would be if the engine failed, or we damaged a float; it had everything except fresh water which was in unlimited supply all around. There were sleeping bags, protective clothing, dehydrated food, medical supplies, tools, emergency radio, rope, matches, torches, everything, in fact, needed for us to be able to look after ourselves.

We took off, wandering home over limitless beauty, landing on several of Bob's other favourite lakes, with me trying my hand at flying with floats. Then we saw people – just four of them portaging two canoes over a bump of land between two lakes. They were only visible when directly below, looking straight down between tall pines at a little splash of red anorak and yellow canoe: then they had gone and alone we flew back to Lac du Bonnet. Bob brought the Cessna in short, right beside their own jetty, and we secured her for the night in the slow ripples of her own reflection.

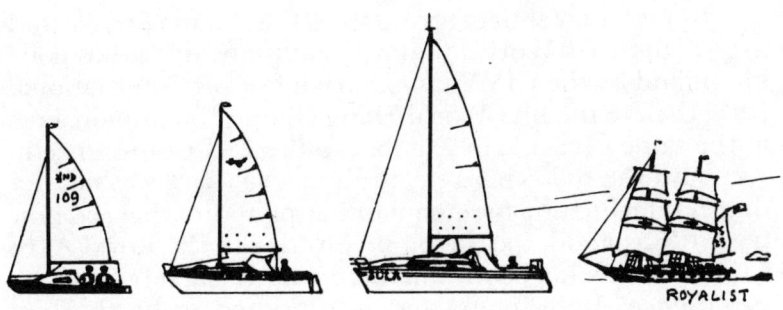

20 · Loss, but many compensations

When Pirat decided to retire from the CIVV hot seat in 1976, I was considered as President-elect and would, if voted in, take office at the Paris meeting in March. But at home my concern to help hang gliding had made my loyalties suspect and a few days before the CIVV meeting the BGA Council withdrew me as CIVV delegate: and if not a delegate I could not become President. Pirat, the FAI Director General Charles Hennecart, and Philip Wills, now BGA President, all wrote to the BGA Chairman asking that the Council reconsider the decision, but it declined to do so. So after almost forty years in which gliding and the BGA had been so much my life there was only one course open to me in protest at such a blow; I resigned.

I have never been a women's lib. fan, having faith that anything I had to offer would be accepted on its merits – in any case there were more interesting things to think about. But looking back dispassionately it did seem that, as a girl, I had had to work somewhat harder for what I wanted. Now I could not help thinking, just a little bit, that if I had been a man the ganging up against me would not have been so determined. But worrying would get me nowhere, and there was no point either in making a public issue of injustice or ceasing to support the gliding I loved. I still saw hang gliding, not as something quite different from gliding, but the cheap and simple end of an expensive and sophisticated sport – rather like dinghies or sailboards, and yachts. There was room, and a great need, for both.

Although now shut off from the BGA, I had been elected to go to the 1976 World Gliding Championships at Räyskälä in Finland by the CIVV, as Chairman of the International Jury: then to the FAI World Hang Gliding Championships in the same capacity, in October to the FAI Conference in Iran and the following year to Rome, so there was neither time nor inclination to remain unhappy about what was now past. It was good, too, being involved with the Royal Aero Club, to which Ian Scott-Hill was bringing stability and long experience. June in Finland is supposed to be the best month for soaring, with high cloudbase and long hours of sunshine; the practice week followed this pattern with hopes that Räyskälä would not turn out to provide yet another example of unexpected and untypical championships weather. But it did just that. As the Opening Ceremony approached clouds gathered and the temperature fell so far that competitors and crews spent the evenings huddled in the saunas for warmth. Then political clouds were added to the meteorological ones. The presence of the South Africans brought forth an objection from the Russians – although they continued to compete – and a threatened withdrawal of financial support by Finland's education ministry.

I had little in the way of protests to deal with; much more of my time seemed to be needed in trying to circumvent political rumblings. There were, as usual, no problems between pilots. When the American Dick Butler, in the lead, badly broke his Glasflugel 604 in an outlanding, and seemed to be out of the Championships, there was no shortage of help: Klaus Holighaus, designer of the rival Nimbus and only 51 points behind Dick in the Championships; Gerhard Waibel, designer of the rival ASW–17, in second place; François Ragot, in sixth place for France; Walter Schneider, designer of the rival LS–3; as well as the Finnish Eiravion repair group, all homed in on the broken glider. With Dick Schreder, Butler's team manager, himself

Loss, but many compensations

designer and constructor of the HP gliders, they worked together all night and, miraculously, got the 604 back in the air for the next day's contest. Gliding's great tradition of friendship in championships was obviously in no danger of demise.

After a brief sunny respite the Räyskälä weather again turned sour, and the last contest day started misty and muggy after a night of continuous heavy rain. No one except the organisers expected a task, but a triangle was set and the gliders set off over the forests and lakes in feeble lift and 3-mile visibility. At one time there were ten of the big 21-m span glass ships circling together in a weak thermal at less than 200 ft, over almost unlandable country. It was a time for steely nerves. No one returned to the downpour that was now Räyskälä, but it was a good day for Britain. RAF Phantom pilot George Lee in his first Worlds deservedly won this weatherbeaten Championships.

I had often been asked whether I minded being on the ground in World Championships while others flew, but I know that I am not a competitive person in terms of wanting to beat other people. Neither do I think that I could summon the courage to go back for more of the tough flying so often demanded in important contests. My will to win surfaces more readily in trying to extract the final shreds of energy from the day's last thermal – or to create, or protect, things I believe in. As team manager I was entirely happy in trying to do my best for the pilots whose own will to win had brought them to the top, and as Chairman of the International Jury to try to take the heat out of problems, so that pilots could concentrate on the flying for which they came. In an odd sort of way I suppose it was the same sort of satisfaction that I got from teaching flying.

After Räyskälä, Teheran for the FAI Conference could not have been more different – hot, dry and lavish. We delegates stayed in different hotels with a fleet of five green buses connecting us to the various Conference halls and receptions. This was fine until we returned home in the late evenings, when it was apparent that the drivers had also found a party. Stimulated, they raced the buses three abreast down the middle of the city's roads. One night I lost count after our driver – in the lead – had jumped seven red traffic lights, to the muttering sound of international prayers. Sadly, it would soon be Iran's turn to pray, but this was unknown to us as we gathered our papers, consolidating another year's FAI work.

The following year, 1977, the Conference was in Rome, and after the working sessions we were given a day out at the local flying club, where I flew a Caproni two-seater glider with Alvaro d'Orleans Borbon, an up-and-coming World Championships pilot. It was delightful floating about in the warm hazy air, but something of a shock when I suddenly realised that I could not remember when I had flown last – I knew it was some months ago, but how many I couldn't recollect. I knew our sailing had taken up almost every weekend, but somehow the urge to get into the air had not intruded into those days on the sea. Back on the ground I started thinking. Time seemed to be accelerating out of control: how could all those months have disappeared without any flying. Had I, in fact, given up flying, or had it given me up – or were those months just a pause to refresh? I loved flying just as much as I had always done, and seemed to be increasingly deeply involved with it, if that were possible. But was my shortage of time real, or was it an excuse? Was it perhaps time I stopped flying – sixty years old was a neat and tidy age to do so: but real enthusiasts continued on, sometimes to seventy or more. I couldn't make up my mind to take any sort of positive decision one way or the other. Lorne had let his licence lapse a few months earlier, though we had neither of us talked at all about giving up flying, as it was a very personal thing to both of us, and neither wanted to influence the other. Our serious sailing had started so as to do something new before we were too old, but to do it properly – our lives in flying had not left us totally ignorant about the need for competence and sound decisions. Without either of us intending, sailing was filling any spare time we had; and unlike gliding, it was something we were doing together, and we were happy doing it.

But all that could come under the heading of excuse, and did not answer the question of whether I should now give up flying, or had actually done so. It was neither a decision or conclusion to be arrived at in a day, or even easily at all. As an instructor I had checked many pilots who had not flown for some time, or were becoming undecided about continuing, so I was well aware of the heartsearching it entailed. I was certainly afraid of increasingly narrowing my abilities, such as they were, by doing only the sort of flying I liked – or was able to afford. It was a downward path that I did not want to follow. As a young, new pilot I had been prepared to try anything that would add to my skills, but in recent years there certainly had been the occasion when I had declined to fly – at the time on grounds of weather or because my instincts said so. But now I was not so sure; maybe I was losing control over my long-time streak of caution and it was now in charge. But mainly I felt jealous of such skills as I possessed and I did not like the idea of losing them gradually and by default, until one day I would break an aircraft through some silly mistake I knew I would not have made five years ago. I had, certainly with some of the luck that every pilot needs, remained free of any but the most trivial aircraft damage for well over forty years, and I did not want to end my flying days as I had seen others end theirs – by kindly advice from some young instructor that perhaps the time had come . . . Maybe I should accept the inevitable; that I had now stopped flying instead of continuing in agonies of doubt as to whether I had or should do so. If nothing else, this would remove the constant feeling of guilt that, instead of doing what I was doing, I ought to go to the airfield and fly. Once the words 'ought to' never existed – if there was a chance of flying nothing would have stopped me, certainly not shortage of time. Probably, because of the enormity of this decision to me, I came to a compromise. I would accept that I had effectively stopped flying to remove the guilty feeling from my mind, but I would not go so far as to say categorically to myself that I would not fly in the future if an interesting opportunity presented itself – and I had time.

Of my involvement I felt no slackening of interest, particularly in the FAI work and in hang gliding: my flying background enabled me to contribute without reservation, and working with people now coming into flying gave me real pleasure. So I had to thank Alvaro – though I still haven't done so – for unknowingly crystallising a crucial moment in my life. I also should have thanked Alvaro's father – the King of Spain's uncle – for a hilarious couple of hours before leaving Rome for home. While packing, my room telephone rang and the receptionist said, 'Madame Welch, your driver is here.' Mystified, as I had not ordered a taxi, I went downstairs to find Alvaro Senior at the desk! We sat in the sun on the terrace talking, until we were joined by an American couple on some package tour who introduced themselves and asked Alvaro what he did. 'I have a little vineyard,' he said plesantly. The American, fascinated, asked under what name the wines were produced.

'd'Orleans.'

'Well, Mr Dorleens, that's interesting, can you tell me where I can buy your wines in the States?'

'Oh, they are not so important as to be exported.'

'Oh, that's bad. Perhaps you would let me know when they come to the States?' He offered his card to Alvaro who thanked him courteously for his interest. 'Well, goodbye Mr Dorleens, Mrs Dorleens' (to me).

'Have a good journey home,' we said, half rising as they left.

It was as well that I had quieted my feelings of guilt about flying or not flying as 1978 was a hectic year. It started with the more comfortable new boat we had promised ourselves for our old age, and for which we had been saving hard. We had finally decided on a Dufour 2800 with a 13 hp diesel, and again we called her *Sula*; our nice Etap *Sula* having been sold to an enthusiast in The Netherlands. After the extensive fitting-out inevitable with any new boat we took our holiday early, sailing 'big' *Sula* to the Isles of Scilly and back. But almost as soon as we were home it was off to France for the World Gliding Championships at Chateauroux in the gently rolling land of the Loire. This time Norwegian Tor Johannessen, who had taken over my CIVV Rules and Championships subcommittee, was to be Chairman of the International Jury, and he asked me if I would go as Deputy Chairman. Since doing things has always been more fun than being someone, I accepted quickly as Tor was a fine colleague to work with, and this was to be the first World Championships with three classes: Open, 15-m span unrestricted, and Standard; also with a 15-m span limit plus restrictions similar to the earlier Standard Class. It was a compromise but looked like being a satisfactory one.

At Chateauroux the weather was good, and it stayed that way. The only problem was that the Director took too much advantage of it, setting such enormous triangles that pilots had to fly for survival rather than speed. Not only did these increase the number of outlandings and retrieve miles, but the pecking order settled into a rut with the same pilots staying at or near the top. After seven days of flying they were tiring, so as diplomatically as possible Tor and I suggested that if the tasks were not quite so long pilots could use a wider range of skills. This helped but not in time to give anyone a chance to dislodge the ever friendly George Lee from his second world win, or Helmut Reichmann from his third – this time flying the exotic carbon-fibre variable-geometry SB–11, built by Akaflieg Braunschweig. The Standard Class was won by Baer Selen, of The Netherlands, in his first Championships, so all three winners were pilots who had become World Champions at their first attempt.

In November the FAI Conference was at Santiago in the Chilean spring. It was a delight to again meet Chilliwilly, who had been aerotowed over the Andes to the 1963 Championships in Argentina, and to stay for the week before the Conference with Rosmarie and Hernan Acuña, who had been at Chateauroux.

It is a strange thing, but I do not think I really appreciated people as people until quite late in life, until I had the chance to trek around the world. In every country I went to I found wonderful friends. In Chile they took me to their homes, their gliding and flying clubs; and up into the foothills of the Andes in search of the hang gliders. We found them in the bright, clean mountain air, their pilots longing to hear news from Europe about flying and how to do it better, while high overhead the condors soared, proud in their mastery.

It was back to Europe again for the 1981 World Gliding Championships. Germany, with its great aviation tradition, has always had a special love for gliding, and so when I was asked to help Tor again with the jury work, and Teddy Stedtfield with the daily Championships magazine, I was very happy. The airfield was to be Paderborn-Haxterberg in north-west Germany, in sight of the Teuterburger Wald hills along which, in 1930, Robert Kronfeld had struggled for the first ever 100 km distance in his Wien. As the cumulus of that long-forgotten summer floated overhead he wondered whether he dared leave the security of the ridge lift to ride those passing clouds. Somehow the dedication of the early pioneers transmitted itself to the Championships in 1981, for the poor weather forced the same need to use every flicker of lift to fight for every kilometre, even to achieve one field further. Led by Fred Weinholtz, himself directly linked to the great days of German gliding, and whose life revolved round his love for flying – and teaching it – Paderborn was a happy Championships. There was only one sad note. Pirat, gliding's friend of thirty-five years, was ill. Every competitor and helper signed a great get-well card and posted it with hope in their hearts. Only later I learnt that he died just a few hours before it arrived, and I thought back to long ago Samedan in the high Alps, where we had first met, to the fun we had in Borovo; and to the many CIVV meetings in Paris where, having finished our work, Pirat, Seff, Piero Morelli, Per Weishaupt from Denmark, and I would drink and talk together at our favourite pavement café. Exactly a year after he died I was again in Paris for the FAI Council meeting and that evening, quite spontaneously, ten of us had dinner together in a tiny French restaurant. We were nine nationalities; Swiss, two British, Australian, Indian, South African, Israeli, Swedish, American, and Dutch, and we enjoyed ourselves and laughed together as is only possible with good friends of long standing. I am sure Pirat was there in spirit.

21 · Messing about with microlights

The last few delightful years of Conference and Championships work had also brought exciting news of yet another stimulus for the little end of aviation – microlights. Grown from hang gliding technology, which in turn had been spurred into life by conventional aviation's unenterprising middle age, it was just as badly needed as hang gliding if anyone with only ordinary money was to be able to fly. Here were real aeroplanes in miniature. For the new pilot a 50-km cross-country flight in such a simple aeroplane would be an adventure, packed with the same need for decisions as for faster flying – plus a few extra for occasional unpremeditated field landings! No airports were necessary, and all aviating could be over open country far from towns or the complications of controlled airspace. The purpose of a microlight for many pilots was, after all, to fly without having to go anywhere.

I could see microlights as a flying which teaches quickly since, with such light, slow aircraft the effects of drift and turbulence are very obvious; just as is the need to avoid having your map blown away in the air. If the pilot made mistakes, as every pilot does, the price of error would be far less than with heavier and faster aeroplanes; so I was just happy that after so many years of aviation becoming more and more complicated, solemn and expensive here was a chance to fly for anyone with energy and enthusiasm.

One thing, at any rate, that I learnt from my own flying, particularly from teaching it, is that it is not only for his own satisfaction that a pilot must decide for himself what to do. If others take the decisions – when to take off and which way to go, or help him by radio anytime he gets lost – he is only partially a pilot. When sooner or later he has to take a vital decision for himself he will not be very good at doing it, as right decisions come only with experience; you have to make mistakes, and learn from them.

By 1980 the world had created around 10,000 skeletal microlight birds, and in Britain some people had formed a Powered Hang Gliding Club; but it did not properly get going. In the autumn a general meeting was called, which I had been asked to chair as an independent arbiter. There, the Club was caused to disappear and be replaced by the British Minimum (later Microlight) Aircraft Association, with Steve Hunt becoming its Chairman. Then in early winter, when I was at the FAI Conference in New Zealand (after a great week with gliding friends sailing and harvesting the orange crop) I was asked to set up a working group to bring these little aircraft under the umbrella of the FAI for their international encouragement. We recommended that a new Committee should be introduced, and the FAI Council agreed to its first formal meeting in Paris in March 1981. So far so good.

Back home the new BMAA was faced with the daunting task of getting itself organised and at the same time negotiating with the Civil Aviation Authority to try to obtain self-regulation. This worked well in gliding and hang gliding and would be just as safe, and enormously cheaper for microlight pilots – who were mainly impecunious – than if expensive licences and fees were imposed. But there was no case for freedom unless these slow, lightweight aeroplanes could be so clearly defined that it was obvious how they differed from conventional ones. This was the first priority at the FAI meeting, and it was Steve Hunt who proposed the definition that was later ratified by the FAI Conference. It ensured that the microlights remained air-

Messing about with microlights

craft of low kinetic energy and therefore basically safe; and it was also accepted by the CAA.

For the previous three years the Authority had turned a blind eye to this developing aviation, which progressed well, far from even a hint of collision with a jumbo jet. Then there was a fatality to a powered hang glider and the CAA sat up and took notice; though before they had decided to become involved the cause of the accident had disappeared because its technical reason, and remedy, was apparent. By this time the BMAA was introducing pilot competence and instructor standards, and schools and clubs were appearing in some numbers. Certainly these were not all good, and certainly there were cowboy elements around; but neither had the early days of any other sort of flying been free from turmoil. I knew that given a little time and responsibility for itself microlight aviation would sort itself out, as had the others. But the CAA did not see it this way and in spite of protests drew these lightweights inexorably into laws made for big, expensive aviation.

Pilots were appalled at the plethora of licences, ratings, restrictions and charges that had suddenly come between them and their flying. Frustration spiralled along with costs, so once again hopes of getting into the air faded, particularly for the young enthusiast without much money. Other countries saw it better. France required proved knowledge of air law, meteorology and navigation, and that the aircraft should be identifiable; the same as our FAI Microlight Committee (CIMA) recommendations. West Germany and Finland had handed regulation over to the Aero Club, while the Americans, ever in support of aviation at all levels, limited flying to daylight, away from controlled airspace – and allowed the pilot to fly at fifteen. All this was sensible since any pilot, licenced or not, has to obey the international rules of the air. The BMAA pilot certificates of competence and the instructor ratings met the special needs of microlights with their varied control systems, yet the CAA still tried to make it fit the existing Navigation Order. Using the work done by the BMAA they added their own charges, and were, in fact, creating a less safe situation by leaving responsibility for carrying out work with the BMAA but keeping the authority. Where national associations had both the responsibility and the authority, as with hang gliding, incompetent or irresponsible operators quickly found themselves policed by their fellows, anxious that their freedom from state legislation should not be taken away. Even a state pilot licence gave little incentive to improve compared to the progressive certificates of hang gliding and gliding – like the Silver and Gold Cs that I worked my way through over the years; pilots like to achieve the higher standards that cannot be bought.

Having always found so much to look forward to I have never, until writing this book, spent much time looking back. Nevertheless, comparisons are not always odious. In aviation most things have happened before and there is no sense in not learning the lessons. Almost all the pioneers were individuals working for the flying they believed in, and if they had been stifled by even a proportion of the legislation that there is today aviation would not have developed to the degree it has. It was only because they, themselves, could decide how to progress that what we have today was achieved. We might not have had the Moth or the Spitfire – or those flights, not always successful – that came lovingly into my teenage diary. If hang gliding had been lumbered with the microlight legislation it would not have survived, let alone produced an industry and top pilots in World Championships. If equivalent legislation charges had been levied in the 1930s when every penny I had went on getting into the air I could not have kept my licence valid and would not have got into ATA. Sometimes, after studying yet another pile of inappropriate requirements I wondered why I did not just retire, and stop fighting for this flying that I did not have the time – or youth – to enjoy myself. The people who wanted to fly microlights could make the objections, go to dreary meetings on summer's only fine day, and argue for freedom. It would be so easy for me to write a single letter of resignation, and I would be free to sail, start painting again, see my neglected friends, and do the many things I now had no time for. But then I would find myself at a little microlight club in a helpful farmer's field, feel again the old love for these new little aeroplanes cavorting in the sunshine, and think of my many friends over the years who had given so much for the flying they loved, and who had helped me to fly. I could not but think back once more to those years when all I wanted to do was to get into the air; and I had been able to do so because I was lucky to be young at a time when aviation was encouraged. Certainly, many of us did not know much and we made plenty of mistakes, but the aircraft of the time were slow and light and we learnt to fly them without doing much damage. Now microlights that were even slower and lighter were having to be fought for every inch of the way to survive in a regulatory structure never intended for this sort of flying. It was by chance that I became involved at the beginning of September 1982 with a microlight race from London to Paris – one of aviation's classic pioneer routes. It was not actually a race but a demanding test of skill and, with 50 km of sea to cross, of courage. A French pilot, Bernard Lamy, was in charge.

The first stage was from Biggin Hill airfield in Kent to Lydd airport on the English side of the Channel. The weather was anticyclonic and the calm sea flecked with gold as one by one sixty-four new pioneers took off to climb away into the haze. From high over the sea, flying with Didier Lamy, Bernard's son, in a Piper Arrow they looked like lonely coloured flies suspended, motionless. On the French side of the water, at Griz Nez lighthouse, observers waited, ready to call the coastguard if anyone fell in the sea. But no one did. Out of that afternoon sky, singly or together, came all sixty-four microlights.

Of course some pilots were not quite on target, and another had to fly holding a loose, hot, silencer with his hand, but the challenge had been met and microlights had come of age. The success – and safety – of this first big competition continued right to the finish at Bagatelle in the centre of Paris; where 67,000 spectators waited to see them arrive.

Unfortunately, at home, the effect of the newly imposed licencing was beginning to bite. Steadily the number of beginners coming into the schools dwindled, frightened off by paperwork and charges they had not expected; and because fewer trained pilots emerged sales of aircraft from microlight's new little industry died away. I felt sad that such great opportunities for enterprise and skill were so inexorably disappearing. But perhaps it was I who was wrong; maybe the enthusiasm for microlights was just a passing whim and in a year's time they would be forgotten. Maybe I had failed to recognise that the golden age of flying was finally over and there was no place for fun in computerised aviation. Thoughts of retiring again floated into my mind. Then at the beginning of December I was invited to Iceland to talk with the Aero Club about microlights and hang gliding, and accepted immediately, as did Bob Harrison, the BHGA training officer.

I had loved Iceland since I had first stayed in that remote country some ten years before in the endless days of summer, though December sounded like endless night. It was not only that I found Iceland infinitely beautiful but the Icelanders' attitude to flying was so much in tune with my own. Faced with weather of unpredictable ferocity sweeping across desolate glaciers and mountains they had a realistic compromise of regulation and competence without stifling enthusiasm. At Akureyri in the far north, where the air traffic controller Huni Snaedal had built a Volksplane and now flew a microlight, we spent an evening with the Aero Club. Just about every glider, hang glider, microlight and aeroplane pilot turned up even though the streets were covered in ice. Next morning we flew back to Reykjavik where it had snowed heavily, and that evening went to the gliding club at the foot of its huge, black soaring ridge with Leifur Magnusson, national gliding champion as well as vice-president of Icelandair. The little road to the club, under two feet of snow, was being dug out by members, but inside their homebuilt clubhouse it was warm, the walls and floor were of polished wood, and ready on a large table was a magnificent buffet supper. It was just their usual winter monthly club meeting, they said.

On our last afternoon it was snowing again, great white squalls flying on a screaming wind, so our hang gliding friends took us into the mountains to walk over the hills from which they fly in summer. We drove up sheep tracks until we slithered to a stop, clambered out into drifts to look over the edges of steep ridges, eyes half closed against the wind-flung snow, and spent hilarious minutes priming a quiescent geyser with soap – to achieve an instant torrent of steam roaring up into the dark cloud racing overhead. Then, enthusiasm renewed, we drove to Keflavik in the last of a pale grey afternoon to fly back that night to our soft green winter.

At home more paper had arrived, more bleak regulatory words that would neither provide the control wanted by authority nor good opportunities to learn and to fly for the pilot. Somehow there had to be simpler legislation which would balance order with opportunity – and not take away the fun. But how to achieve it? Maybe, after all, this was still not quite the moment to retire.

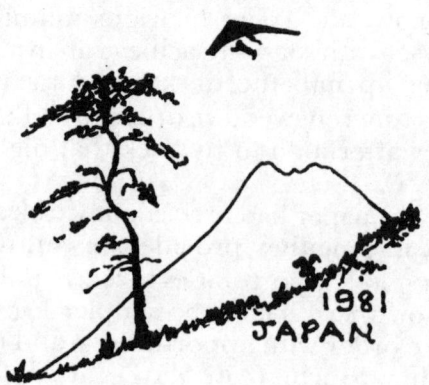

22 · The dream remembered

It was after our 1981 summer holiday sailing *Sula* in thick windless haze or gales, followed by a quick assault on the overgrown garden, that I set off for Japan. In early October the World Hang Gliding Championships would start at Beppu, on the island of Kyushu, and at the end of the month there would be the FAI Conference in Tokyo.

For both events, and for me, it was first time in the Far East, although Lorne had been many times on business. I flew there over the North Pole in twenty-four hours of sunshine, being met by Asahi Miyahara, now seventy-seven, and Director of the Championships, which put my thoughts of retiring to shame; most of the competitors would be less than a third of his age.

The Japan Aeronautic Association, as well as the city of Beppu, had put enormous efforts into their first Championships, even to carving a landing field out of the mountains, but the weather was not an ally, even in this land of bright cherry blossom, and relatively little serious contest flying was possible due to rain, wind and a passing typhoon. This was a pity as hang gliders, such as the Comet, were capable of soaring quite long distances and their pilots were longing to do so; though neither at Grenoble nor at Beppu had this chance come. But with international hang gliding

The dream remembered

now six years old it was a great meeting of friends who had managed to find their way across the world. The only protest we had was on a technicality between the Japanese team and the Japanese organisers, which caused me as Jury chairman to pay attention to what I was doing, as I did not think either would relish the idea of being in the wrong in front of foreigners!

Between the Championships and the Conference I stayed with the Miyaharas in Tokyo. It was a wonderful opportunity to learn about this extraordinary country, where each day produced something unexpected, and see more of Asahi and Hisako Miyahara other than at aviation meetings – and a few hours at our home which Asahi spent lying on his tummy studying the newts in the garden pond! In 1924, aged twenty, he had been dispatched by his father to Glasgow University, and then apprenticed to the Westland Aircraft Company at Yeovil, returning to Japan in 1929 to work for Mitsubishi. From England he had brought a copy of *Mrs Beeton's English Cook Book* for his wife-to-be, and the bound volume of *Flight* magazine for 1927 – my bedtime reading in 1981. As an aircraft designer back in Japan one of his early achievements had been a remarkably modern-looking motor glider.

The two weeks went fast exploring the coast and the mountains, looking at a large hang gliding school which changed to a ski resort in November – on artificial snow if the real thing was being dilatory – and a gliding club struggling to survive on a tough mountaintop 5000 ft up, because there is no spare flat land. Now I knew why the Japanese never did well in World Gliding; the geography of their country makes it almost impossible to do any serious cross-country soaring. It, plus the weather, was the same problem that afflicted Iceland. We also went to the Yamaha factory at Hammamatsu where I gave a lecture about microlights. I was not asked to do it until the night before at an excellent but formal Japanese dinner party, in which I had the usual European problem of what to do with my legs when after about three minutes I got pins and needles from sitting on them. I wrote some notes in bed – my room had three kingsize double beds so it took me some time to decide which one to use. Next morning the lecture was to start at 10.30 but half an hour earlier I was told that the President of Yamaha wanted to meet me. So Mrs Miyahara, myself, and four Department heads presented ourselves in his office, shook hands, bowed and sat on the edges of chairs to drink our green tea. Then I reckoned we would bow again and depart for the lecture – except that when the President asked if I liked music I said that I did. Beaming, he brought his quadraphonics into top gear to pour piano music over us. When it stopped he said it had been composed and was played by an eleven-year-old Japanese child, and started to play more and tell me about the Yamaha Music Foundation and how wonderful it was for children to learn to love classical music. By now we were 15 minutes late for my lecture and I was becoming concerned in case the patient audience were muttering about Europeans who did not get up in the morning. One of the Department heads was trying to say we ought to go, without leaving his head on the block, but the President told him with an airy wave that music was more important. At 11.30, we left the presence for the lecture; so forty employees had done nothing for an hour – 1 man week. But no one seemed to be worried.

And so my Japanese education continued, from high-speed trains to lunch at the club of the disinherited hereditary peers of Japan, on the thirty-fourth floor of a modern building, with five full-sized billiard tables and golf on the carpet; from unbelievably hot communal baths, to being the only European – and woman – invited to a luncheon party given by the Mayor of Beppu to his Prince; and learning to cook Japanese food in the Miyahara kitchen with Asahi in an apron enjoying himself as chief chef.

All too soon I had to move into the hotel in the centre of Tokyo for the FAI Conference, with the kind Miyaharas coming every day to see how I was getting on. For me it was a very special Conference because in Tokyo I was awarded the Gold Air Medal of the FAI, which later in England I was honoured to receive from Prince Charles. It was a long time from cycling to Biggin Hill and Croydon, leaning on the fence at the Lancashire Aero Club in the early thirties, or soloing the Moth in a daze of excitement at Barnstable, or the delight of flying my first Spitfire. Looking back it was difficult to see how one thing had led to another. Certainly and for sure, I had been lucky.

But there was no time to reminisce, for as soon as I had returned over the Pole, looking with fascination at 1500 miles of fractured and crumpled ice, wondering how any explorer and his dog team could ever find how to cross those miles of black water leads without satellite assistance, I found I had been given a berth on the training brig *Royalist* in just three days time. This I could not miss. The previous November I had spent a week on *Royalist* as crew, and would never forget racing up Cornwall's Falmouth estuary to shelter from a pursuing gale, with cloud and curtains of fine rain over the hills, and nothing visible, except the working boats dredging for oysters under sail. It was as though we were 150 years back in time. Then, topsails still set, and water rushing along the lee gunwale, we sped up the narrow reach between tree-covered hills past the *King Harry* ferry, and round the corner to shelter. Moored safely, we listened to the 50-knot wind tearing through the ancient oak trees and rattling our rigging. So how could I possibly put off any chance of sailing again on *Royalist*? So away went the Conference clothes and files of paper, and into a duffle bag went sweaters and oilies. Dear Lorne drove me to Gosport, without any complaint of another week feeding the cats!

On board I was delighted to see the same friendly faces, but very surprised to find myself as a watch officer – and the only woman on board; I could only conclude that they must have been very short of November volunteers. But there was no time to have doubts as the cadets were already arriving and I had forgotten which rope did what. I was given forward port watch with five boys – the youngest not quite thirteen and the eldest just fifteen, and they were fun to work with. Because of the weather – (one day I must write a book with that title) – our Captain, Francis Drake, reckoned that we would only get anywhere in the week if we night-sailed to Alderney in the Channel Isles. He didn't particularly like doing this so early in any trip with very young cadets, so eased the problem by cutting night watches from four to two hours. My watch started at 2 a.m., and I was glad of my years as a gliding instructor, because in the middle of the Channel, in the middle of the winter, and the middle of the night I found myself teaching these five young lads, for the first time in their lives, to steer a big square-rigged ship; with two of them only nose high to the compass. When they were not doing their 15–20 minutes on the wheel zigzagging earnestly across the dark sea and bringing noisy rattles from protesting topsails, I had them as look-outs with instructions to point out any vessel they saw. Port look-out seemed to be somewhat quiet so I asked what he could see. 'Nothing, Ma'am. The liferafts are in he way.' 'Move aft,' I said, 'so you can see around them.' There was a short pause, then 'Ooh, Ma'am, there's a big ship.' He had at last spotted the lights of a cross-channel ferry passing us about half a mile distant like a battalion of Christmas trees. His eyes widened because it was suddenly all real to him, and not second-hand, like television.

Three days later we had a superb sail home to Gosport, a force 7 south-westerly on our quarter and *Royalist* storming along at 11 knots. It was a good feeling to be crew on such a day, the ship alive with the power of the wind driving her hour after hour through a grey-white sea under the racing clouds. This was the same sea-wind that I had loved as a child when it threw sparkles of spray from the thundering waves and rustled sand through the dune grass. In some way this wind had always been part of my life so that after a few calm days I began to feel that something was missing.

But it was not only the gusting wind blowing out of a blue-washed sky that was so exciting. I also loved the light-fingered wind that moved *Sula* through water so smooth that her reflection was barely disturbed or that would gently lift a glider up the face of a great mountain, although with a breath so soft it seemed to have no power at all. Unexpectedly, a long-forgotten flight came fresh to my mind. It was at Davos in the summer of 1949 when I had been lent an elderly, slow Spahlinger S–18 and was one morning launched from the little airfield near the lake. The electric winch, no more successful than at Samedan, had given me 450 ft with which to reach the towering tree-covered north-east face below the 3000-ft Jacobshorn. The soft summer air

The dream remembered

was drifting on to it but the steep slope was still in its shadow, waiting for the sun. I flew straight and carefully for the patterned cover of dark firs and larches, as any lift would be close among them, and ready to turn along the mountain at the last moment. My gull-winged glider was still slowly sinking as I banked, and only when I straightened up again did the gentle wind help me – and fill the cockpit with the resin scent of the trees. The spur of this mountain was barely a kilometre long and beyond it, both ways, there would be only sinking air; so unless I could climb higher between turns I would not even have height to get back to the small airfield. Afraid to go too far I turned back along the slope early – and tightly – so as not to leave the upward drift of the air as it threaded its way around the treetops. One pine was dead, a skeleton still reaching for the light. Just holding my gains I flew back along the steep face, still in cool shadow, to turn over a break in the trees. In it was a narrow path with two climbers, packs on their backs, heads down, unseeing as they made their laborious way to the sunny snowfield above.

On the next beat the dead tree was a few feet lower and as I turned again, more hopefully now, a hawk rose just ahead. It swung round quickly to soar above me and effortlessly led the way. I tried to follow but the hawk had already gone, gaining in minutes the few hundred feet that I might win only after a breath-holding hour. Then I met the climbers again. They had stopped at another break in the trees and were looking out over the valley as I flew silently above them; turned, flew back, and called, 'Hallo.' They looked around then suddenly up, and waved. Then they were behind me and I flew on alone, exploring and exploiting every breath of gently rising air, unconscious of myself in the cramped little cockpit, my wings a part of me. Again I turned, to float quietly back the way I had come over the scented trees, in no hurry to go anywhere except to reach the sunlit air above me. Imperceptibly, each time I passed it, the dead pine was just a little lower and looking up I could see more of the bright snow which filled the mountaintop. Above me, in the dazzling sun sailed the hawk, looking down at the poor big bird still in shadow. Then from the last of the trees came the two climbers and I watched them take off their anoraks and look up for the glider. But they had reached the sun before me and I was floating along just below them, looking up. Suddenly they saw me again, close, and smiled a welcome. Then I too came above the trees and flew into a great surge of lift as the gentle wind met its mother the sun, and for another hour played in the bright sky, just happy to fly.

Index

A licence, 13, 15
A 15 (glider), 169
Acuña, Hernan & Rosmarie, 227
Aeronca C-3, 15
Air Transport Auxiliary (ATA), 39, 51, 73
Airacobra, 65
Airspeed Ferry, 4
Akaflieg Braunschweig, 227
Akureyri, 234
Albacore, 57
Alvis Silver Eagle, 13
Allen, Charlie, 11
Anglo-German Fellowship, 18
Angoulême, 136, 182
Anson; Avro Anson, 46, 64
Argosy, A.W., 10
Armstrong, Stan & Pat, 167
Army Gliding Club, 94
ASK-14, 188
ASW-17, 222
Aston Down aerodrome, 41, 44
Atalanta, A.W., 9
Audax, Hawker, 48
Auster G-AGVJ, 70, 72, 74, 77, 93
Avro 504K, 10
Avro Avian, 9, 10, 16
Avro Cadet, 12
Ax, Goran, 204
Aydogan, Ziya, 176

BA Swallow, 16
BAC Drone, 16
Bailey, Lady, 9
Bailey, Robert, 216
ball lightning, 63
Barker, Michael, 174
Barnard, Alec, 35
Barnard, Mollie, 13
Barnstaple aerodrome, 13
Battle of Britain, 39
Battle, Ken, 213
Beaverette, 77
Benemann, Jochen, 33
Beppu, 236, 238

Berchtesgaden, 33
Bergel, Hugh, 20
BGA, 73, 84, 91, 164, 172, 188,
BGA Instructors' Panel, 92, 116, 173, 175
BGA No.1 Test Group, 75, 80, 117, 188
BHGA, 208
Bickel, Dick, 208
Bickley (Kent), 6
Biggin Hill, 11, 233
Birrfeld, 81
Bishop Hill, 28
Blackburn Bluebird, 10
Blackburn Shark, 57
Blackwell, Anne, 73
Blanchard, Paul, 113
Blanik (glider), 154
Blenheim, Bristol, 66
Blunt, Vernon, 71
BMAA, 230
Bocian (glider), 150
Bonney, Mrs, 9
Borbon, Alvaro d'Orleans Jnr, 224
Borbon, Alvaro d'Orleans Snr, 226
Borovo, 104
Bosham Sailing Club, 198
Boyd, R.J., 14
Brabazon, Lord, 86
Bramcote (RNAS), 77
Breguet 901, 111
Bridson, Doug, 174
Brooklands, 14, 25, 50
Brooks, Peter, 90
Brown, Charles, 81
Brown, Don, 93
Brunt, Professor, 73
BSA car, 13
Buckingham, Horace, 75
Buckland (Surrey), 31
Bulldog, Bristol, 11
Burg Feuerstein, 190
Burgess, Gerry, 154
Burton, George, 112, 175, 178, 180, 185, 194, 203
Butler, Dick, 222

Index

Butler, Lois, 54
Butzweilerhof, 133

Cambridge University GC, 70
Camphill (Derbyshire), 98, 108
Carlisle aerodrome, 23, 37, 43
Caproni 2-seater, 224
Cardiff, John, 194
Carr, Johnny, 216
Cartry, Jean Pierre, 169, 204
Cessna 180 amphibian, 217
Cessna Skylane, 184
Chateauroux, 226
Chattis Hill, 59
Chesapeake, Vought-Sikorsky, 57
Chillywilly, 154, 227
Chipmunk, DH 168, 171
Cierva C19, 10
Cijan, Boris, 104
CIMA, 231
CIVL, 213, 214
CIVV (CVSM), 2, 164, 170, 172, 183, 197, 212, 221, 226
Clarke, Alan, 88
Claudi, Ron, 86
Cobham's Circus, 4
College of Aeronautical Engineering, 25
Colley Hill (Surrey), 25, 31
Collins, Eric, 70
Compte, René, 136
Cornwall, N., 7, 64
Corsellis, Tim, 61
Corston, Chris, 209
Cowdray Park, 57
Cronk, David, 212
Crossley, Winnie, 54
Croydon aerodrome, 7, 11
Cub, Taylor, 16
Cunnison, Margaret, 40

D-36 (glider), 171
Dagling (glider), 19, 31
Dalrymple, Hon. A.W.H., 75
Dan Air, 113
Davies, S. Kenneth, 199
Davis, Duncan, 16
Davos, 120, 152, 240
Deane-Drummond, Tony, 98, 112, 124, 156
Delafield, John, 180, 203, 204
Delore, Terry, 213
Delta Silver, 216

Derby & Lancs GC, 25, 78, 93, 108, 167, 174
Desoutter, 10
Dewsbury, Jack, 26
Diamonds (gliding), 127, 135, 140, 149
Dimitrovski, 105
Dominie (Rapide), 54
Douglas, Elizabeth (Liz), 70, 120, 171, 179
Douglas, Graham, 30, 32, 37, 43, 68, 75
Douglas, Vivien, 69
Drake, Francis, 239
Drew, Harold, 174
Dufour 2800, 226
Duncan, Ricky, 214
Dunstable, 3, 18, 20, 30, 93
Dunstan, Don, 202
Duperier, Bernard, 212
Dupont, Steve, 181
Dutton, George, 57
Dziurzynski, Adam, 149

Eagle, Slingsby, 108, 122
Eckersley, Peter, 13
Edelweiss (glider), 169, 170
Ehrat, Emil, 177
Eiravion, 222
Elke, Udo, 176
Elliott's of Newbury, 75, 83, 93, 124
Ellis, Noel, 174
Empire Exhibition (1938), 22
Etap 22, 198
Evans, Tom, 169
Everitt, John, 174, 175

Fabre, Henri, 216
FAI, 2, 96, 128, 168, 192, 199, 210, 212, 223, 236, 238
Fairchild Argus, 61, 64
Fairey Battle, 37, 38, 50
Fairey Fulmar, 63
Falcon III (glider), 30
Falke (glider), 22
Falke, Scheibe, 187, 189, 191
Farnell, Diane, 47
Ferry Pilots Notes, 57
Fieseler Storch, 75, 84
Findlay, Max, 16
Firth, John, 205
Fitchett, Bernard, 194, 203
Fleet Finch, 156
Fleet Thrush, 157
Flight magazine, 16, 122

Index

Foka (glider), 134, 170
Forbes, Jock, 83, 96, 98
Foster, Frank, 98, 121, 124
Fox, John, 70
Fox Moth, DH, 41
Fripp, Mike, 177
Furlong, John, 153, 154

Gawron (aeroplane), 149
Gehriger, Pirat, 84, 96, 97, 104, 105, 172, 190, 202, 221, 228
Gladiator, Gloster, 63
Glider in the Sky (film), 175
Gold Air medal, 238
Gold C, 182, 232
Goodwood (W. Hampnett), 56
Gower, Pauline, 40, 48
Goodhart, Nick, 97, 112, 121, 124, 132, 153, 175, 180, 194
Goodhart, Tony, 98, 112, 124
Gossamer Albatross, 208
Gough, Andy, 180
Grant, Malcolm, 61
Greaves, Con, 180
Greig, Donald, 26, 32, 83, 84, 87, 89
Grenoble, 215
Grey, C.G., 16
Grosse, Hans Werner, 204, 205
Grunau Baby (glider), 22, 25, 37, 70
Gull IV (glider), 81, 83, 120
Gunn, Jimmy, 17

Haase, Ernst-Gunther, 127, 132, 134
Haggard, Roy, 212
Hamble aerodrome, 48, 57, 63
Harrison, Bob, 234
Harvard (T6), 48
Hatfield aerodrome, 23, 25, 39
Hawker Demon, 11, 48
Hawker Hurricane, 24, 38, 50
Hayes, James, 34
Haynes, Stan, 72
Heinkel III, 38
Hennecart, Charles, 221
Henry, François, 170
Heracles, HP, 11
Herringshaw, Stan, 61
Hervey, Tim, 2, 19, 20
Herzogen Cecilie, 198
Hess, Rudolf, 35
Hill, Bridget, 61
Hill, Roy, 216
Hinkler, Bert, 9

Himberger, Sepp, 211, 214
Hiscox, Dudley, 91
HKS (glider), 127, 132
Hookings, Gordon, 200
Holighaus, Klaus, 204, 222
Hope, Captain, 9
Hornet Moth, DH, 23
Horsfield, Brenda, 95
Hossinger, Rolf, 134
Howe, Gillian, 174
HP14 (glider), 178, 222
Humphries, Sebert, 20
Hunt, Martin, 209
Hunt, Steve, 230
Huth, Heinz, 156
Hutter Libelle (glider), 170
Huish, 27

Iceland, 234
Iggulden, Jack, 160
Imperial College GC, 73, 76, 93, 95
Ince, David, 98
Inkpen (Berks), 27, 98
Innes, David, 175, 197
Innes, Sheila, 197
Irving, Frank, 93, 117, 125, 128, 171, 174, 176, 179, 188, 198
Ivans, Bill, 122, 123, 183

Jacobs, Hans, 70, 75
Jameson, Don, 61
Japan Aeronautic Association, 236
Jaskolka (glider), 140
Jefferson, Bryan, 126
Jeffs, Jimmy, 11
Jenkins, Roy, 169
Johannessen, Tor, 226, 228
Johnson, Amy, 41
Johnson, Dick, 101, 160, 203
Jones, Davy, 202
Jones, Dukinfield, 32, 37
Junkers G-31, G-38, 11
Junin, Argentina, 152

K1, 'Crabpot', 103
Ka6 (glider), 159, 174
Kahn, Wally, 93
Karran, Jack, 76, 94
Kemsley Trust, 98
Kendall, Hugh, 73, 80, 103
Kenley aerodrome, 71, 73
Kent, 'Pops', 85
Kepka, Franzicek, 205

Index

Kerr, Sir John, 206
Kestrel (glider), 194
Kidder, R., 110
Kinder, Frank, 95
King, Alison, 57
King Kite (glider), 71
Kirby Kadet, 31
Kirbymoorside, 28
Kite (glider), 26
KLM, 11
Knes, Marian, 104, 106
Komac, Bozidar, 109
Kosava (glider), 105, 109, 123
Kössen, 210
Kronfeld, Robert, 228
Kuntz, Rolf, 159
Kunz, Seff, 190, 228
Kupchanko, Dean, 213

La Ferté Allais, 136
Lamy, Bernard, 233
Lamy, Didier, 233
Lancashire Aero Club, 12
Lancaster, Captain, 9
Land Rover, 83, 86
Lasch, Helli, 99
Lasham, airfield, 94, 111, 112, 120, 164, 174, 180, 189
Lauber, Eddie, 87
Lee, George, 223, 227
Leleu, L., 11
Leopard Moth, DH, 47
Lerwell, George, 153
Leszno, Poland, 124, 144, 175
Lever, Graham, 61
Licher, Lloyd, 184, 193, 207
Licher, Rose Marie, 184, 207
Lilienthal medal, 206
List, Ivor, 17
Lomax, David, 175
London GC, 93 (*see* Dunstable)
Long Mynd, 73
'Low and Slow', 192
LS-3 (glider), 222
Lundy aerodrome, 14

McBroom, Geoff, 216
MacCready, Paul, 96, 101, 122, 208
Mackay, Bob, 208
McKelvie, L.J., 174
Magister, Miles, 43
Magnusson, Leifur, 234

Makula, Edward, 161
Mallet, Pete, 83, 96
Marfa, 180, 182
Marfa Air, 184
Marwell aerodrome, 59
Maurer, Siegbert, 86
Meads, Basil, 98
Meise (Olympia), 75
Messenger, Ken, 210
Messerschmitt, 39
Meteor (glider), 105
microlights, 229
Midwood, Harry, 125
Miles Hawk, 10
Miles Master, 49
Miller, Richard, 192
Minimoa, 72
Minton, Paul, 117
Mistral (wind), 122
Mix, Wolf, 196
Miyahara, Asahi, 210, 236
Miyahara, Hisako, 237
Moffat, George, 186, 204, 205
Mollison, James, 9
Monospar, 10
Morelli, Piero, 173, 184
Morison, Walter, 73
Moswey III (glider), 80
Moth, DH, 10, 14, 32
Moth Major, 15
Moth Minor, 32
Motor Falke, 187
Moyes, Bill, 212
Mü 13 (glider), 72
Mucha Standard, 140
Murray, Bill, 27

Nash, J.W., 14
Nathan, Lord, 91
Neaves, Roger, 175
Neilan, John, 93
Nicholson, E.Q., 87
Nicholson, Kit, 20, 83, 87
Nietlispach, Hans, 109, 203
Nillson, Billy, 96
Nimbus (glider), 222
Norbert, Walter, 184

Oda, Isamu, 154
Olympia 419, 124
Olympia (Meise), 75, 77, 83, 93
Orao (glider), 104
Orebro, Sweden, 96

Index

Orsi, Adele, 203
OSTIV, 132
Ottewill, Group Captain, 166, 171
Owen, Ken, 122, 125
Owen, Sel, 200
Oxford, Airspeed, 56, 57

Paderborn, W. Germany, 228
Pearson, Ted, 161
Pechaud, M., 131
Percival Gull, 10
Perranporth airfield, 61, 135
Persson, Pelle, 87, 123
Petroczy, Gyorgi, 169
Petterson, Ake, 197
Phoebus (glider), 170, 182
Phoenix (glider), 134, 170
Pierre, Gerard, 99, 110
Piggott, Derek, 113, 189
Po-2 (aeroplane), 104, 108, 148
Popular Flying, 8
Powell, Brian, 31
Poynter, Dan, 212, 214
Pressland, Joy, 85
Prestwick aerodrome, 46
Prince Charles, HRH, 238
Prince Philip, HRH, 112, 125, 168
Pronzati, A., 197
Purves, Bob, 217
Purves, Doreen, 217
Puss Moth, DH, 9, 10, 11, 41, 61
Pye radio, 98

Queen Bee, DH, 41

RAFGSA, 164
Ragot, François, 204
Rain, Zvonimir, 109
Rambaut, Christopher, 66
Räyskälä, Finland, 222
Redhill Flying Club/airfield, 32, 37, 70
Regan, Nick, 209
Reichmann, Helmut, 186, 205, 227
Reid, Derek, 93
Renner, Ingo, 205
Rhönadler (glider), 70
Rhönsperber (glider), 26
Rice, Jack, 77
Riddell, Chris, 171
RJ-5 (glider), 101
Rogallo, Francis, 192
Rolls Silver Shadow, 176
Rossfeld Ski Hut, 33

Roundway Down, Wilts., 26, 28
Royal Aero Club, 16, 199
Royal Aeronautical Society, 9
Royalist, TS, 239
Ryan, John, 182
Rzeszow, Poland, 148

Sailplane and Gliding, 193
St Auban, France, 123
St Yan, France, 121
Salzgitter, W. Germany, 71
Samedan, Switzerland, 83, 136
SB–11 (glider), 227
Saradic, A., 122
Schachenmann, Max, 87, 99
Scheibe SF-25 Falke, 187, 189, 191
Scherne, Werner, 212
Schneider, Walter, 222
Schreder, Dick, 133, 161, 222
Schubert, Ekkehard, 177
Schweizer 2-25, 109
Scott, Peter, 112
Scott-Hill, Ian, 222
Scottish Gliding Union, 28
Seafire, 66
Selen, Baer, 227
SG-38 (glider), 81
Shaw, Widdle, 20, 22, 24, 26, 30
Shimek, Anton, 104
Shirach, Baldur von, 34
SHK (glider), 178
Silloth aerodrome, 53
Silver C, 22, 72, 93, 183, 232
Simons, Martin, 203
Sir Winston Churchill, TS, 198
Skua, Blackburn, 62
Sky (glider), 98
Skylark 2 (glider), 117
Skylark 3 (glider), 121, 124, 134, 154
Slater, Doc, 71
Slingsby, Fred, 27, 83, 98, 168, 178
Slingsby Cadet (glider), 76
Slingsby Dart (glider), 177
Slingsby Swallow (glider), 173
Smith, A.J., 178, 194, 197
Smith, Bill, 48
Smith, Stan, 110
Smith, Victor, 9
South Cerney aerodrome, 44, 164, 192
Sowery, John, 97
Spahlinger S-18, 240
Spahlinger S-25, 77, 78
Spartan 3-seater, 10

Spence, Warren, 155
Spitfire, 56–68
Spooner, Reggie, 216
Spyr (glider), 109
Stafford-Allen, Ray, 174
Standard Class, 112, 124, 170, 177, 192
Stearman, 156, 160
Stedtfeld, Teddy, 228
Steinbach brothers, 214
Stephenson, Geoffrey, 26, 32, 98, 121
Stolle, Gerd, 192
Sula 1, 179
Sula 2, 198
Sula 3, 226, 236
Surrey GC, 32, 70, 76, 93
Sutton Bank, 27
Swordfish, Fairey, 56

T21 (glider), 92, 113, 119
T49 Capstan, 119
Tabart, Tony, 204
Taylor Cub, 15
Ternhill aerodrome, 50
Teuterburger Wald, 228
Thevenot, Gerard, 216
Thompson, Tommy, 200
Thorburn, Andrew, 29
Tiger Moth, DH, 15, 40, 70, 77, 94
Tipsy Nipper, 117
Torcross, S. Devon, 8, 36
Torrell, Ara, 87
Trotter, Hugo, 76, 95
Tunnicliffe, Charles, 12
Tysowsze, Poland, 146

Unimog, 101
Unterberg, Tyrol, 211

V-1 flying bomb, 69
V-2 rocket, 69
Vanguard, Standard, 98, 124
Varkosi, Lagos, 196
Vauxhall 30/98, 15
Vrsac, Yugoslavia, 194, 197

Waibel, Gerhard, 222
Waikerie, S. Australia, 200
Wallington, C.E. 'Wally', 112, 125, 167, 168, 171, 200, 205

Walrus, Vickers-Supermarine, 57, 58
Watson, Dick, 95
Weihe (glider), 73, 83, 84, 93, 96, 104
Weinholtz, Fred, 228
Weishaupt, Per, 228
Welch, Jan (Janet), 120, 171, 172, 179
Welch, Lorne, 73, 74, 78, 85, 93, 96, 98, 103, 104, 113, 120, 124, 140, 154, 174, 179, 216, 224
Wellington, Vickers, 66
Westerley Nimrod, 179, 197
White Waltham (WW), 48, 66, 117
Whitlam, Gough, 204
Whittle, Monty, 11
Wien (glider), 226
Wijewardine, Ray, 71
Wilkes, Peter, 86
Wilkinson, Ken, 193
Williamson, Henry, 8, 12, 13, 125
Williamson, John (John Willy), 125, 154, 157, 175, 177, 203
Wills, Kitty, 87, 98, 130
Wills, Philip, 20, 83, 87, 93, 96, 98, 102, 121, 124, 130, 164, 172, 175, 199, 221
Wills, W.D. & H.O., 165, 173
Wilson, Colin, 13
Wingfield, Charles, 83, 85
Witek, Adam, 132
Wödl, Harro, 178, 184
Wojtulanis, Barbara, 59
Woodford aerodrome, 12
Woods, Jimmy, 10
Worthington, George, 216
Wright, Wilbur, 216
Wrigley, Barbara, 171
Wroblewski, Jan, 170, 197
Würth, Hans, 77

Yak 18 (aeroplane), 149
Yamaha, 238
Youell, George, 13

Zabiello, Irene, 125, 139, 175
Zabiello, Roman, 125, 139, 175
Zar, Poland, 149
Zefir (glider), 134, 161
Zegels, Bert, 204
Zermatt, Switzerland, 77